Plate Tectonics: A Comprehensive Introduction

Plate Tectonics: A Comprehensive Introduction

Edited by
Fernando Morrison

目 **Larsen & Keller**
www.larsen-keller.com

Plate Tectonics: A Comprehensive Introduction
Edited by Fernando Morrison
ISBN: 978-1-63549-226-2 (Hardback)

Larsen & Keller

Published by Larsen and Keller Education,
5 Penn Plaza,
19th Floor,
New York, NY 10001, USA

Cataloging-in-Publication Data

Plate tectonics : a comprehensive introduction / edited by Fernando Morrison.
      p. cm.
Includes bibliographical references and index.
ISBN 978-1-63549-226-2
1. Plate tectonics. 2. Geology, Structural. 3. Geodynamics.
I. Morrison, Fernando.
QE511.4 .P53 2017
551.136--dc23

This book contains information obtained from authentic and highly regarded sources. All chapters are published with permission under the Creative Commons Attribution Share Alike License or equivalent. A wide variety of references are listed. Permissions and sources are indicated; for detailed attributions, please refer to the permissions page. Reasonable efforts have been made to publish reliable data and information, but the authors, editors and publisher cannot assume any responsibility for the vailidity of all materials or the consequences of their use.

Trademark Notice: All trademarks used herein are the property of their respective owners. The use of any trademark in this text does not vest in the author or publisher any trademark ownership rights in such trademarks, nor does the use of such trademarks imply any affiliation with or endorsement of this book by such owners.

The publisher's policy is to use permanent paper from mills that operate a sustainable forestry policy. Furthermore, the publisher ensures that the text paper and cover boards used have met acceptable environmental accreditation standards.

Printed and bound in the United States of America.

For more information regarding Larsen and Keller Education and its products, please visit the publisher's website www.larsen-keller.com

Table of Contents

Preface

Plate tectonics is the theory which deals with the study of movements of the seven large plates and other smaller plates that compose the lithosphere of Earth. It is crucial in the study of the geographical movement and evolution of the Earth's landmass as well as for studying and forecasting volcanic and seismic activities. This book unfolds the innovative aspects of the area which will be crucial for the holistic understanding of the subject matter. The topics covered in this extensive text deal with the core subjects of plate tectonics. This textbook is meant for students who are looking for an elaborate reference text on this subject area.

A foreword of all chapters of the book is provided below:

Chapter 1 - The Earth has seven large plates and several smaller plates in its lithosphere. The theory describing the motion of these plates is known as plate tectonics. The chapter on plate tectonics offers an insightful focus, keeping in mind the subject matter; **Chapter 2** - In order to develop a firm understanding of tectonics it is very important to understand concepts such as thrust tectonics, strike-slip tectonics, salt tectonics, neotectonics etc. Thrust tectonics is particularly concerned with the thickening of the crust whereas strike-slip tectonics is concerned with the structures formed by levels of displacement within the crust. This section is an overview of the subject matter incorporating all the major aspects of tectonics; **Chapter 3** - Pacific Plate is the tectonic plate that exists underneath the Pacific Ocean. The North American Plate similarly covers North America, Greenland and Cuba. The other major tectonic plates covered within this section are Eurasian Plate, African Plate, Antarctic Plate, Australian Plate, South American Plate, Scotia Plate and Indian plate. This chapter elucidates the main tectonic plates; **Chapter 4** - The outer shell of any planet is termed as the crust. It is majorly formed by igneous processes and it occupies less than 1% of the Earth's volume. Some other aspects elucidated in the section are oceanic crust, asthenosphere, mantle and lithosphere. The chapter offers an insightful focus on the topic and incorporates its major aspects; **Chapter 5** - Continental drifting is the shifting of the continents of the Earth and it has been occurring over a period of millions of years. The other key concepts of plate tectonics that have been explained in this chapter are divergent boundary, continental collision, plate reconstruction, crustal recycling etc. The section strategically encompasses and incorporates the major components and key concepts of plate tectonics, providing a complete understanding; **Chapter 6** - A convergent boundary is where two or more tectonic plates move towards one another and then clash into one another. Subduction, obduction, orogeny and transform fault are some of the aspects explained in the following text. The aspects discussed in the chapter are of vital importance, and provide a better understanding of tectonic plates; **Chapter 7** - Mid-ocean ridge is an underwater mountain system; this system is formed by plate tectonics. The various mid-ocean ridges are Mid-Atlantic Ridge, South American-Antarctic Ridge and Central Indian Ridge. The following section also focuses on topics such as seafloor spreading, oceanic trench, passive margin, volcanic passive margin etc; **Chapter 8** - A supercontinent is the collected landmass of almost all of the Earth's continents. An ancient supercontinent mentioned in the text is Gondwana. This section serves as a source to understand all the major supercontinents, and provides a better understanding on the subject matter; **Chapter 9** - An earthquake occurs because of the energy suddenly released by the Earth's crust. Earthquakes cause immense damage to life and to property. The topics covered in this section are seismotectonics, intraplate earthquake and interplate earthquake. The topics discussed in the chapter are of great importance to broaden the existing knowledge on Earthquakes.

At the end, I would like to thank all the people associated with this book devoting their precious time and providing their valuable contributions to this book. I would also like to express my gratitude to my fellow colleagues who encouraged me throughout the process.

Editor

Introduction to Plate Tectonics

The Earth has seven large plates and several smaller plates in its lithosphere. The theory describing the motion of these plates is known as plate tectonics. The chapter on plate tectonics offers an insightful focus, keeping in mind the subject matter.

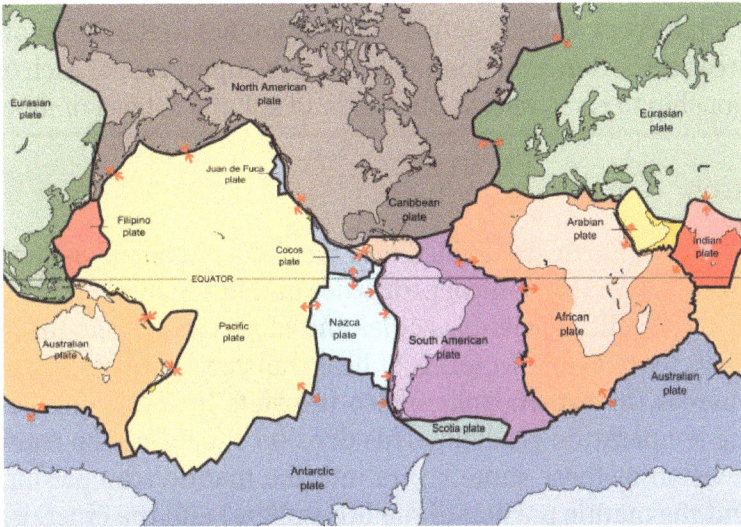

The tectonic plates of the world were mapped in the second half of the 20th century.

Plate tectonics is a scientific theory describing the large-scale motion of 7 large plates and the movements of a larger number of smaller plates of the Earth's lithosphere, over the last 100's of millions of years. The theoretical model builds on the concept of continental drift developed during the first few decades of the 20th century. The geoscientific community accepted plate-tectonic theory after seafloor spreading was validated in the late 1950s and early 1960s.

The lithosphere, which is the rigid outermost shell of a planet (the crust and upper mantle), is broken up into tectonic plates. The Earth's lithosphere is composed of seven or eight major plates (depending on how they are defined) and many minor plates. Where the plates meet, their relative motion determines the type of boundary: convergent, divergent, or transform. Earthquakes, volcanic activity, mountain-building, and oceanic trench formation occur along these plate boundaries. The relative movement of the plates typically ranges from zero to 100 mm annually.

Tectonic plates are composed of oceanic lithosphere and thicker continental lithosphere, each topped by its own kind of crust. Along convergent boundaries, subduction carries plates into the mantle; the material lost is roughly balanced by the formation of new (oceanic) crust along divergent margins by seafloor spreading. In this way, the total surface of the lithosphere remains the same. This prediction of plate tectonics is also referred to as the conveyor belt principle. Earlier theories (that still have some supporters) propose gradual shrinking (contraction) or gradual expansion of the globe.

Tectonic plates are able to move because the Earth's lithosphere has greater strength than the underlying asthenosphere. Lateral density variations in the mantle result in convection. Plate movement is thought to be driven by a combination of the motion of the seafloor away from the spreading ridge (due to variations in topography and density of the crust, which result in differences in gravitational forces) and drag, with downward suction, at the subduction zones. Another explanation lies in the different forces generated by tidal forces of the Sun and Moon. The relative importance of each of these factors and their relationship to each other is unclear, and still the subject of much debate.

Key Principles

The outer layers of the Earth are divided into the lithosphere and asthenosphere. This is based on differences in mechanical properties and in the method for the transfer of heat. Mechanically, the lithosphere is cooler and more rigid, while the asthenosphere is hotter and flows more easily. In terms of heat transfer, the lithosphere loses heat by conduction, whereas the asthenosphere also transfers heat by convection and has a nearly adiabatic temperature gradient. This division should not be confused with the *chemical* subdivision of these same layers into the mantle (comprising both the asthenosphere and the mantle portion of the lithosphere) and the crust: a given piece of mantle may be part of the lithosphere or the asthenosphere at different times depending on its temperature and pressure.

The key principle of plate tectonics is that the lithosphere exists as separate and distinct *tectonic plates*, which ride on the fluid-like (visco-elastic solid) asthenosphere. Plate motions range up to a typical 10–40 mm/year (Mid-Atlantic Ridge; about as fast as fingernails grow), to about 160 mm/year (Nazca Plate; about as fast as hair grows). The driving mechanism behind this movement is described below.

Tectonic lithosphere plates consist of lithospheric mantle overlain by either or both of two types of crustal material: oceanic crust (in older texts called *sima* from silicon and magnesium) and continental crust (*sial* from silicon and aluminium). Average oceanic lithosphere is typically 100 km (62 mi) thick; its thickness is a function of its age: as time passes, it conductively cools and subjacent cooling mantle is added to its base. Because it is formed at mid-ocean ridges and spreads outwards, its thickness is therefore a function of its distance from the mid-ocean ridge where it was formed. For a typical distance that oceanic lithosphere must travel before being subducted, the thickness varies from about 6 km (4 mi) thick at mid-ocean ridges to greater than 100 km (62 mi) at subduction zones; for shorter or longer distances, the subduction zone (and therefore also the mean) thickness becomes smaller or larger, respectively. Continental lithosphere is typically ~200 km thick, though this varies considerably between basins, mountain ranges, and stable cratonic interiors of continents. The two types of crust also differ in thickness, with continental crust being considerably thicker than oceanic (35 km vs. 6 km).

The location where two plates meet is called a *plate boundary*. Plate boundaries are commonly associated with geological events such as earthquakes and the creation of topographic features such as mountains, volcanoes, mid-ocean ridges, and oceanic trenches. The majority of the world's active volcanoes occur along plate boundaries, with the Pacific Plate's Ring of Fire being the most active and widely known today. These boundaries are discussed in further detail below. Some volcanoes occur in the interiors of plates, and these have been variously attributed to internal plate deformation and to mantle plumes.

As explained above, tectonic plates may include continental crust or oceanic crust, and most plates contain both. For example, the African Plate includes the continent and parts of the floor of the Atlantic and Indian Oceans. The distinction between oceanic crust and continental crust is based on their modes of formation. Oceanic crust is formed at sea-floor spreading centers, and continental crust is formed through arc volcanism and accretion of terranes through tectonic processes, though some of these terranes may contain ophiolite sequences, which are pieces of oceanic crust considered to be part of the continent when they exit the standard cycle of formation and spreading centers and subduction beneath continents. Oceanic crust is also denser than continental crust owing to their different compositions. Oceanic crust is denser because it has less silicon and more heavier elements ("mafic") than continental crust ("felsic"). As a result of this density stratification, oceanic crust generally lies below sea level (for example most of the Pacific Plate), while continental crust buoyantly projects above sea level.

Types of Plate Boundaries

Three types of plate boundaries exist, with a fourth, mixed type, characterized by the way the plates move relative to each other. They are associated with different types of surface phenomena. The different types of plate boundaries are:

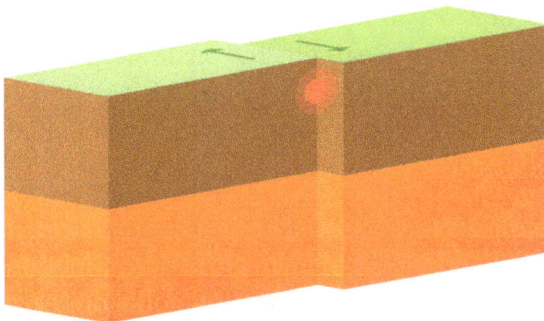

Transform boundary

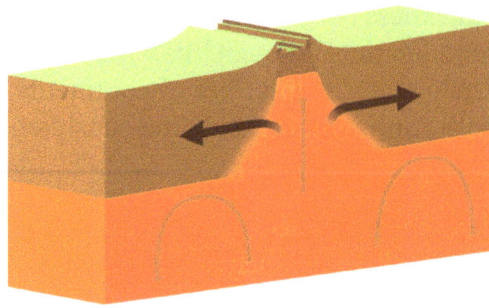

Divergent boundary

Convergent boundary

1. *Transform boundaries (Conservative)* occur where two lithospheric plates slide, or perhaps more accurately, grind past each other along transform faults, where plates are neither created nor destroyed. The relative motion of the two plates is either sinistral (left side toward the observer) or dextral (right side toward the observer). Transform faults occur across a spreading center. Strong earthquakes can occur along a fault. The San Andreas Fault in California is an example of a transform boundary exhibiting dextral motion.

2. *Divergent boundaries (Constructive)* occur where two plates slide apart from each other. At zones of ocean-to-ocean rifting, divergent boundaries form by seafloor spreading, allowing for the formation of new ocean basin. As the ocean plate splits, the ridge forms at the spreading center, the ocean basin expands, and finally, the plate area increases causing many small volcanoes and/or shallow earthquakes. At zones of continent-to-continent rifting, divergent boundaries may cause new ocean basin to form as the continent splits, spreads, the central rift collapses, and ocean fills the basin. Active zones of Mid-ocean ridges (e.g., Mid-Atlantic Ridge and East Pacific Rise), and continent-to-continent rifting (such as Africa's East African Rift and Valley, Red Sea) are examples of divergent boundaries.

3. *Convergent boundaries (Destructive)* (or *active margins*) occur where two plates slide toward each other to form either a subduction zone (one plate moving underneath the other) or a continental collision. At zones of ocean-to-continent subduction (e.g. the Andes mountain range in South America, and the

Cascade Mountains in Western United States), the dense oceanic lithosphere plunges beneath the less dense continent. Earthquakes trace the path of the downward-moving plate as it descends into asthenosphere, a trench forms, and as the subducted plate is heated it releases volatiles, mostly water from hydrous minerals, into the surrounding mantle. The addition of water lowers the melting point of the mantle material above the subducting slab, causing it to melt. The magma that results typically leads to volcanism. At zones of ocean-to-ocean subduction (e.g. Aleutian islands, Mariana Islands, and the Japanese island arc), older, cooler, denser crust slips beneath less dense crust. This causes earthquakes and a deep trench to form in an arc shape. The upper mantle of the subducted plate then heats and magma rises to form curving chains of volcanic islands. Deep marine trenches are typically associated with subduction zones, and the basins that develop along the active boundary are often called "foreland basins". Closure of ocean basins can occur at continent-to-continent boundaries (e.g., Himalayas and Alps): collision between masses of granitic continental lithosphere; neither mass is subducted; plate edges are compressed, folded, uplifted.

4. *Plate boundary zones* occur where the effects of the interactions are unclear, and the boundaries, usually occurring along a broad belt, are not well defined and may show various types of movements in different episodes.

Driving Forces of Plate Motion

Plate motion based on Global Positioning System (GPS) satellite data from NASA JPL. The vectors show direction and magnitude of motion.

It has generally been accepted that tectonic plates are able to move because of the relative density of oceanic lithosphere and the relative weakness of the asthenosphere. Dissipation of heat from the mantle is acknowledged to be the original source of the energy required to drive plate tectonics through convection or large scale upwelling and doming. The current view, though still a matter of some debate, asserts that as a consequence, a powerful source of plate motion is generated due to the excess density of the oceanic lithosphere sinking in subduction zones. When the new crust forms at mid-ocean ridges, this oceanic lithosphere is initially less dense than the underlying asthenosphere, but it becomes denser with age as it conductively cools and thickens. The greater density of old lithosphere relative to the underlying asthenosphere allows it to sink into the deep mantle at subduction zones, providing most of the driving force for plate movement. The weakness of the asthenosphere allows the tectonic plates to move easily towards a subduction zone. Although subduction is thought to be the strongest force driving plate motions, it cannot be the only force since there are plates such as the North American Plate which are moving, yet are nowhere being subducted. The same is true for the enormous Eurasian Plate. The sources of plate motion are a matter of intensive research and discussion among scientists. One of the main points is that the kinematic pattern of the movement itself should be separated clearly from the possible geodynamic mechanism that is invoked as the driving force of the observed movement, as some patterns may be explained by more than one mechanism. In short, the driving forces advocated at the moment can be divided into three categories based on the relationship to the movement: mantle dynamics related, gravity related (mostly secondary forces).

Driving Forces Related to Mantle Dynamics

For much of the last quarter century, the leading theory of the driving force behind tectonic plate motions envisaged large scale convection currents in the upper mantle, which can be transmitted through the asthenosphere. This theory was launched by Arthur Holmes and some forerunners in the 1930s and was immediately recognized as the solution for the acceptance of the theory as originally discussed in the papers of Alfred Wegener in the early years of the century. However, despite its acceptance, it was long debated in the scientific community because the leading ("fixist") theory still envisaged a static Earth without moving continents up until the major breakthroughs of the early sixties.

Two- and three-dimensional imaging of Earth's interior (seismic tomography) shows a varying lateral density distribution throughout the mantle. Such density variations can be material (from rock chemistry), mineral (from variations in mineral structures), or thermal (through thermal expansion and contraction from heat energy). The manifestation of this varying lateral density is mantle convection from buoyancy forces.

How mantle convection directly and indirectly relates to plate motion is a matter of ongoing study and discussion in geodynamics. Somehow, this energy must be transferred to the lithosphere for tectonic plates to move. There are essentially two main types of forces that are thought to influence plate motion: friction and gravity.

- Basal drag (friction): Plate motion driven by friction between the convection currents in the asthenosphere and the more rigid overlying lithosphere.

- Slab suction (gravity): Plate motion driven by local convection currents that exert a downward pull on plates in subduction zones at ocean trenches. Slab suction may occur in a geodynamic setting where basal tractions continue to act on the plate as it dives into the mantle (although perhaps to a greater extent acting on both the under and upper side of the slab).

Lately, the convection theory has been much debated as modern techniques based on 3D seismic tomography still fail to recognize these predicted large scale convection cells. Therefore, alternative views have been proposed:

In the theory of plume tectonics developed during the 1990s, a modified concept of mantle convection currents is used. It asserts that super plumes rise from the deeper mantle and are the drivers or substitutes of the major convection cells. These ideas, which find their roots in the early 1930s with the so-called "fixistic" ideas of the European and Russian Earth Science Schools, find resonance in the modern theories which envisage hot spots/mantle plumes which remain fixed and are overridden by oceanic and continental lithosphere plates over time and leave their traces in the geological record (though these phenomena are not invoked as real driving mechanisms, but rather as modulators). Modern theories that continue building on the older mantle doming concepts and see plate movements as a secondary phenomena are beyond the scope of this page and are discussed elsewhere (for example on the plume tectonics page).

Another theory is that the mantle flows neither in cells nor large plumes but rather as a series of channels just below the Earth's crust, which then provide basal friction to the lithosphere. This theory, called "surge tectonics", became quite popular in geophysics and geodynamics during the 1980s and 1990s. Recent research, based on three-dimensional computer modeling, suggests that plate geometry is governed by a feedback between mantle convection patterns and the strength of the lithosphere.

Driving Forces Related to Gravity

Forces related to gravity are usually invoked as secondary phenomena within the framework of a more general driving mechanism such as the various forms of mantle dynamics described above.

Gravitational sliding away from a spreading ridge: According to many authors, plate motion is driven by the higher elevation of plates at ocean ridges. As oceanic lithosphere is formed at spreading ridges from hot mantle material, it gradually cools and thickens with age (and thus adds distance from the ridge). Cool oceanic lithosphere is significantly denser than the hot mantle material from which it is derived and so with increasing thickness it gradually subsides into the mantle to compensate the greater load. The result is a slight lateral incline with increased distance from the ridge axis.

This force is regarded as a secondary force and is often referred to as "ridge push". This is a misnomer as nothing is "pushing" horizontally and tensional features are dominant along ridges. It is more accurate to refer to this mechanism as gravitational sliding as variable topography across the totality of the plate can vary considerably and the topography of spreading ridges is only the most prominent feature. Other mechanisms generating this gravitational secondary force include flexural bulging of the lithosphere before it dives underneath an adjacent plate which produces a clear topographical feature that can offset, or at least affect, the influence of topographical ocean ridges, and mantle plumes and hot spots, which are postulated to impinge on the underside of tectonic plates.

Slab-pull: Current scientific opinion is that the asthenosphere is insufficiently competent or rigid to directly cause motion by friction along the base of the lithosphere. Slab pull is therefore most widely thought to be the greatest force acting on the plates. In this current understanding, plate motion is mostly driven by the weight of cold, dense plates sinking into the mantle at trenches. Recent models indicate that trench suction plays an important role as well. However, as the North American Plate is nowhere being subducted, yet it is in motion presents a problem. The same holds for the African, Eurasian, and Antarctic plates.

Gravitational sliding away from mantle doming: According to older theories, one of the driving mechanisms of the plates is the existence of large scale asthenosphere/mantle domes which cause the gravitational sliding of lithosphere plates away from them. This gravitational sliding represents a secondary phenomenon of this basically vertically oriented mechanism. This can act on various scales, from the small scale of one island arc up to the larger scale of an entire ocean basin.

Driving Forces Related to Earth Rotation

Alfred Wegener, being a meteorologist, had proposed tidal forces and pole flight force as the main driving mechanisms behind continental drift; however, these forces were considered far too small to cause continental motion as the concept then was of continents plowing through oceanic crust. Therefore, Wegener later changed his position and asserted that convection currents are the main driving force of plate tectonics in the last edition of his book in 1929.

However, in the plate tectonics context (accepted since the seafloor spreading proposals of Heezen, Hess, Dietz, Morley, Vine, and Matthews during the early 1960s), the oceanic crust is suggested to be in motion *with* the continents which caused the proposals related to Earth rotation to be reconsidered. In more recent literature, these driving forces are:

1. Tidal drag due to the gravitational force the Moon (and the Sun) exerts on the crust of the Earth

2. Global deformation of the geoid due to small displacements of rotational pole with respect to the Earth's crust;

3. Other smaller deformation effects of the crust due to wobbles and spin movements of the Earth rotation on a smaller time scale.

Forces that are small and generally negligible are:

1. The Coriolis force

2. The centrifugal force, which is treated as a slight modification of gravity

For these mechanisms to be overall valid, systematic relationships should exist all over the globe between the orientation and kinematics of deformation and the geographical latitudinal and longitudinal grid of the Earth itself. Ironically, these systematic relations studies in the second half of the nineteenth century and the first half of the twentieth century underline exactly the opposite: that the plates had not moved in time, that the deformation grid was fixed with respect to the Earth equator and axis, and that gravitational driving forces were generally acting vertically and caused only local horizontal movements (the so-called pre-plate tectonic, "fixist theories"). Later studies (discussed below on this page), therefore, invoked many of the relationships recognized during this pre-plate tectonics period to support their theories.

Of the many forces discussed in this paragraph, tidal force is still highly debated and defended as a possible principle driving force of plate tectonics. The other forces are only used in global geodynamic models not using plate tectonics concepts (therefore beyond the discussions treated in this section) or proposed as minor modulations within the overall plate tectonics model.

In 1973, George W. Moore of the USGS and R. C. Bostrom presented evidence for a general westward drift of the Earth's lithosphere with respect to the mantle. He concluded that tidal forces (the tidal lag or "friction") caused by the Earth's rotation and the forces acting upon it by the Moon are a driving force for plate tectonics. As the Earth spins eastward beneath the moon, the moon's gravity ever so slightly pulls the Earth's surface layer back westward, just as proposed by Alfred Wegener. In a more recent 2006 study, scientists reviewed and advocated these earlier proposed ideas. It has also been suggested recently in Lovett (2006) that this observation may also explain why Venus and Mars have no plate tectonics, as Venus has no moon and Mars' moons are too small to have significant tidal effects on the planet. In a recent paper, it was suggested that, on the other hand, it can easily be observed that many plates are moving north and eastward, and that the dominantly westward motion of the Pacific ocean basins derives simply from the eastward bias of the Pacific spreading center (which is not a predicted manifestation of such lunar forces). In the same paper the authors admit, however, that relative to the lower mantle, there is a slight westward component in the motions of all the plates. They demonstrated though that the westward drift, seen only for the past 30 Ma, is attributed to the increased dominance of the steadily growing and accelerating Pacific plate. The debate is still open.

Relative Significance of Each Driving Force Mechanism

The vector of a plate's motion is a function of all the forces acting on the plate; however, therein lies the problem regarding the degree to which each process contributes to the overall motion of each tectonic plate.

The diversity of geodynamic settings and the properties of each plate result from the impact of the various processes actively driving each individual plate. One method of dealing with this problem is to consider the relative rate at which each plate is moving as well as the evidence related to the significance of each process to the overall driving force on the plate.

One of the most significant correlations discovered to date is that lithospheric plates attached to downgoing (subducting) plates move much faster than plates not attached to subducting plates. The Pacific plate, for instance, is essentially surrounded by zones of subduction (the so-called Ring of Fire) and moves much faster than the plates of the Atlantic basin, which are attached (perhaps one could say 'welded') to adjacent continents instead of subducting plates. It is thus thought that forces associated with the downgoing plate (slab pull and slab suction) are the driving forces which determine the motion of plates, except for those plates which are not being subducted. The driving forces of plate motion continue to be active subjects of on-going research within geophysics and tectonophysics.

Development of the Theory

Summary

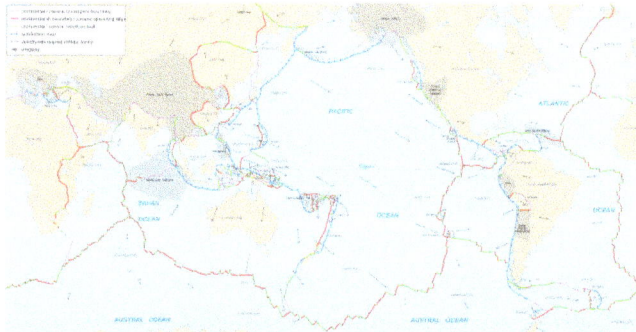

Detailed map showing the tectonic plates with their movement vectors.

In line with other previous and contemporaneous proposals, in 1912 the meteorologist Alfred Wegener amply described what he called continental drift, expanded in his 1915 book *The Origin of Continents and Oceans* and the scientific debate started that would end up fifty years later in the theory of plate tectonics. Starting from the idea (also expressed by his forerunners) that the present continents once formed a single land mass (which was called Pangea later on) that drifted apart, thus releasing the continents from the Earth's mantle and likening them to "icebergs" of low density granite floating on a sea of denser basalt. Supporting evidence for the idea came from the dove-tailing outlines of South America's east coast and Africa's west coast, and from the matching of the rock formations along these edges. Confirmation of their previous contiguous nature also came from the fossil plants *Glossopteris* and *Gangamopteris*, and the therapsid or mammal-like reptile *Lystrosaurus*, all widely distributed over South America, Africa, Antarctica, India and Australia. The evidence for such an erstwhile joining of these continents was patent to field geologists working in the southern hemisphere. The South African Alex du Toit put together a mass of such information in his 1937 publication *Our Wandering Continents*, and went further than Wegener in recognising the strong links between the Gondwana fragments.

But without detailed evidence and a force sufficient to drive the movement, the theory was not generally accepted: the Earth might have a solid crust and mantle and a liquid core, but there seemed to be no way that portions of the crust could move around. Distinguished scientists, such as Harold Jeffreys and Charles Schuchert, were outspoken critics of continental drift.

Despite much opposition, the view of continental drift gained support and a lively debate started between "drifters" or "mobilists" (proponents of the theory) and "fixists" (opponents). During the 1920s, 1930s and 1940s, the former reached important milestones proposing that convection currents might have driven the plate movements, and that spreading may have occurred below the sea within the oceanic crust. Concepts close to the elements now incorporated in plate tectonics were proposed by geophysicists and geologists (both fixists and mobilists) like Vening-Meinesz, Holmes, and Umbgrove.

One of the first pieces of geophysical evidence that was used to support the movement of lithospheric plates came from paleomagnetism. This is based on the fact that rocks of different ages show a variable magnetic field direction, evidenced by studies since the mid–nineteenth century. The magnetic north and south poles reverse through time, and, especially important in paleotectonic studies, the relative position of the magnetic north pole varies through time. Initially, during the first half of the twentieth century, the latter phenomenon was explained by introducing what was called "polar wander" , i.e., it was assumed that the north pole location had been shifting through time. An alternative explanation, though, was that the continents had moved (shifted and rotated) relative to the north pole, and each continent, in fact, shows its own "polar wander path". During the late 1950s it was successfully shown on two occasions that these data could show the validity of continental drift: by Keith Runcorn in a paper in 1956 , and by Warren Carey in a symposium held in March 1956.

The second piece of evidence in support of continental drift came during the late 1950s and early 60s from data on the bathymetry of the deep ocean floors and the nature of the oceanic crust such as magnetic properties and, more generally, with the development of marine geology which gave evidence for the association of seafloor spreading along the mid-oceanic ridges and magnetic field reversals, published between 1959 and 1963 by Heezen, Dietz, Hess, Mason, Vine & Matthews, and Morley.

Simultaneous advances in early seismic imaging techniques in and around Wadati-Benioff zones along the trenches bounding many continental margins, together with many other geophysical (e.g. gravimetric) and geological observations, showed how the oceanic crust could disappear into the mantle, providing the mechanism to balance the extension of the ocean basins with shortening along its margins.

All this evidence, both from the ocean floor and from the continental margins, made it clear around 1965 that continental drift was feasible and the theory of plate tectonics, which was defined in a series of papers between 1965 and 1967, was born, with all its extraordinary explanatory and predictive power. The theory revolutionized the Earth sciences, explaining a diverse range of geological phenomena and their implications in other studies such as paleogeography and paleobiology.

Continental Drift

In the late 19th and early 20th centuries, geologists assumed that the Earth's major features were fixed, and that most geologic features such as basin development and mountain ranges could be explained by vertical crustal movement, described in what is called the geosynclinal theory. Generally, this was placed in the context of a contracting planet Earth due to heat loss in the course of a relatively short geological time.

Alfred Wegener in Greenland in the winter of 1912-13.

It was observed as early as 1596 that the opposite coasts of the Atlantic Ocean—or, more precisely, the edges of the continental shelves—have similar shapes and seem to have once fitted together.

Since that time many theories were proposed to explain this apparent complementarity, but the assumption of a solid Earth made these various proposals difficult to accept.

The discovery of radioactivity and its associated heating properties in 1895 prompted a re-examination of the apparent age of the Earth. This had previously been estimated by its cooling rate and assumption the Earth's surface radiated like a black body. Those calculations had implied that, even if it started at red heat, the Earth would have dropped to its present temperature in a few tens of millions of years. Armed with the knowledge of a new heat source, scientists realized that the Earth would be much older, and that its core was still sufficiently hot to be liquid.

By 1915, after having published a first article in 1912, Alfred Wegener was making serious arguments for the idea of continental drift in the first edition of *The Origin of Continents and Oceans*. In that book (re-issued in four successive editions up to the final one in 1936), he noted how the east coast of South America and the west coast of Africa looked as if they were once attached. Wegener was not the first to note this (Abraham Ortelius, Antonio Snider-Pellegrini, Eduard Suess, Roberto Mantovani and Frank Bursley Taylor preceded him just to mention a few), but he was the first to marshal significant fossil and paleo-topographical and climatological evidence to support this simple observation (and was supported in this by researchers such as Alex du Toit). Furthermore, when the rock strata of the margins of separate continents are very similar it suggests that these rocks were formed in the same way, implying that they were joined initially. For instance, parts of Scotland and Ireland contain rocks very similar to those found in Newfoundland and New Brunswick. Furthermore, the Caledonian Mountains of Europe and parts of the Appalachian Mountains of North America are very similar in structure and lithology.

However, his ideas were not taken seriously by many geologists, who pointed out that there was no apparent mechanism for continental drift. Specifically, they did not see how continental rock could plow through the much denser rock that makes up oceanic crust. Wegener could not explain the force that drove continental drift, and his vindication did not come until after his death in 1930.

Floating Continents, Paleomagnetism, and Seismicity Zones

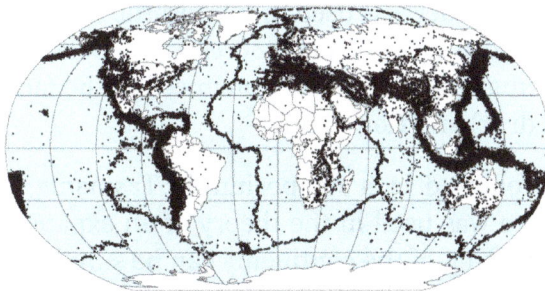

Preliminary Determination of Epicenters
358,214 Events, 1963 - 1998

Global earthquake epicenters, 1963–1998

As it was observed early that although granite existed on continents, seafloor seemed to be composed of denser basalt, the prevailing concept during the first half of the twentieth century was that there were two types of crust, named "sial" (continental type crust) and "sima" (oceanic type crust). Furthermore, it was supposed that a static shell of strata was present under the continents. It therefore looked apparent that a layer of basalt (sial) underlies the continental rocks.

However, based on abnormalities in plumb line deflection by the Andes in Peru, Pierre Bouguer had deduced that less-dense mountains must have a downward projection into the denser layer underneath. The concept that mountains had "roots" was confirmed by George B. Airy a hundred years later, during study of Himalayan gravitation, and seismic studies detected corresponding density variations. Therefore, by the mid-1950s, the question remained unresolved as to whether mountain roots were clenched in surrounding basalt or were floating on it like an iceberg.

During the 20th century, improvements in and greater use of seismic instruments such as seismographs enabled scientists to learn that earthquakes tend to be concentrated in specific areas, most notably along the oceanic trenches and spreading ridges. By the late 1920s, seismologists were beginning to identify several prominent earthquake zones parallel to the trenches that typically were inclined 40–60° from the horizontal and extended several hundred kilometers into the Earth. These zones later became known as Wadati-Benioff zones, or simply Benioff zones, in honor of the seismologists who first recognized them, Kiyoo Wadati of Japan and Hugo Benioff of the United States. The study of global seismicity greatly advanced in the 1960s with the establishment of the Worldwide Standardized Seismograph Network (WWSSN) to monitor the compliance of the 1963 treaty banning above-ground testing of nuclear weapons. The much improved data from the WWSSN instruments allowed seismologists to map precisely the zones of earthquake concentration worldwide.

Meanwhile, debates developed around the phenomena of polar wander. Since the early debates of continental drift, scientists had discussed and used evidence that polar drift had occurred because continents seemed to have moved through different climatic zones during the past. Furthermore, paleomagnetic data had shown that the magnetic pole had also shifted during time. Reasoning in an opposite way, the continents might have shifted and rotated, while the pole remained relatively fixed. The first time the evidence of magnetic polar wander was used to support the movements of continents was in a paper by Keith Runcorn in 1956, and successive papers by him and his students Ted Irving (who was actually the first to be convinced of the fact that paleomagnetism supported continental drift) and Ken Creer.

This was immediately followed by a symposium in Tasmania in March 1956. In this symposium, the evidence was used in the theory of an expansion of the global crust. In this hypothesis the shifting of the continents can be simply explained by a large increase in size of the Earth since its formation. However, this was unsatisfactory because its supporters could offer no convincing mechanism to produce a significant expansion of the Earth. Certainly there is no evidence that the moon has expanded in the past 3 billion years; other work would soon show that the evidence was equally in support of continental drift on a globe with a stable radius.

During the thirties up to the late fifties, works by Vening-Meinesz, Holmes, Umbgrove, and numerous others outlined concepts that were close or nearly identical to modern plate tectonics theory. In particular, the English geologist Arthur Holmes proposed in 1920 that plate junctions might lie beneath the sea, and in 1928 that convection currents within the mantle might be the driving force. Often, these contributions are forgotten because:

- At the time, continental drift was not accepted.

- Some of these ideas were discussed in the context of abandoned fixistic ideas of a deforming globe without continental drift or an expanding Earth.

- They were published during an episode of extreme political and economic instability that hampered scientific communication.

- Many were published by European scientists and at first not mentioned or given little credit in the papers on sea floor spreading published by the American researchers in the 1960s.

Mid-Oceanic Ridge Spreading and Convection

In 1947, a team of scientists led by Maurice Ewing utilizing the Woods Hole Oceanographic Institution's research vessel *Atlantis* and an array of instruments, confirmed the existence of a rise in the central Atlantic Ocean, and found that the floor of the seabed beneath the layer of sediments consisted of basalt, not the granite which is the main constituent of continents. They also found that the oceanic crust was much thinner than continental crust. All these new findings raised important and intriguing questions.

The new data that had been collected on the ocean basins also showed particular characteristics regarding the bathymetry. One of the major outcomes of these datasets was that all along the globe, a system of mid-oceanic ridges was detected. An important conclusion was that along this system,' new ocean floor was being created, which led to the concept of the "Great Global Rift". This was described in the crucial paper of Bruce Heezen (1960), which would trigger a real revolution in thinking. A profound consequence of seafloor spreading is that new crust was, and still is, being continually created along the oceanic ridges. Therefore, Heezen advocated the so-called "expanding Earth" hypothesis of S. Warren Carey. So, still the question remained: how can new crust be continuously added along the oceanic ridges without increasing the size of the Earth? In reality, this question had been solved already by numerous scientists during the forties and the fifties, like Arthur Holmes, Vening-Meinesz, Coates and many others: The crust in excess disappeared along what were called the oceanic trenches, where so-called "subduction" occurred. Therefore, when various scientists during the early sixties started to reason on the data at their disposal regarding the ocean floor, the pieces of the theory quickly fell into place.

The question particularly intrigued Harry Hammond Hess, a Princeton University geologist and a Naval Reserve Rear Admiral, and Robert S. Dietz, a scientist with the

U.S. Coast and Geodetic Survey who first coined the term *seafloor spreading*. Dietz and Hess (the former published the same idea one year earlier in *Nature*, but priority belongs to Hess who had already distributed an unpublished manuscript of his 1962 article by 1960) were among the small handful who really understood the broad implications of sea floor spreading and how it would eventually agree with the, at that time, unconventional and unaccepted ideas of continental drift and the elegant and mobilistic models proposed by previous workers like Holmes.

In the same year, Robert R. Coats of the U.S. Geological Survey described the main features of island arc subduction in the Aleutian Islands. His paper, though little noted (and even ridiculed) at the time, has since been called "seminal" and "prescient". In reality, it actually shows that the work by the European scientists on island arcs and mountain belts performed and published during the 1930s up until the 1950s was applied and appreciated also in the United States.

If the Earth's crust was expanding along the oceanic ridges, Hess and Dietz reasoned like Holmes and others before them, it must be shrinking elsewhere. Hess followed Heezen, suggesting that new oceanic crust continuously spreads away from the ridges in a conveyor belt–like motion. And, using the mobilistic concepts developed before, he correctly concluded that many millions of years later, the oceanic crust eventually descends along the continental margins where oceanic trenches – very deep, narrow canyons – are formed, e.g. along the rim of the Pacific Ocean basin. The important step Hess made was that convection currents would be the driving force in this process, arriving at the same conclusions as Holmes had decades before with the only difference that the thinning of the ocean crust was performed using Heezen's mechanism of spreading along the ridges. Hess therefore concluded that the Atlantic Ocean was expanding while the Pacific Ocean was shrinking. As old oceanic crust is "consumed" in the trenches (like Holmes and others, he thought this was done by thickening of the continental lithosphere, not, as now understood, by underthrusting at a larger scale of the oceanic crust itself into the mantle), new magma rises and erupts along the spreading ridges to form new crust. In effect, the ocean basins are perpetually being "recycled," with the creation of new crust and the destruction of old oceanic lithosphere occurring simultaneously. Thus, the new mobilistic concepts neatly explained why the Earth does not get bigger with sea floor spreading, why there is so little sediment accumulation on the ocean floor, and why oceanic rocks are much younger than continental rocks.

Magnetic Striping

Beginning in the 1950s, scientists like Victor Vacquier, using magnetic instruments (magnetometers) adapted from airborne devices developed during World War II to detect submarines, began recognizing odd magnetic variations across the ocean floor. This finding, though unexpected, was not entirely surprising because it was known that basalt—the iron-rich, volcanic rock making up the ocean floor—contains a strongly magnetic mineral (magnetite) and can locally distort compass readings. This distortion was recognized by Icelandic mariners as early as the late 18th century. More important, because the presence of magnetite gives the basalt measurable magnetic properties, these newly discovered magnetic variations provided another means to study the deep ocean floor. When newly formed rock cools, such magnetic materials recorded the Earth's magnetic field at the time.

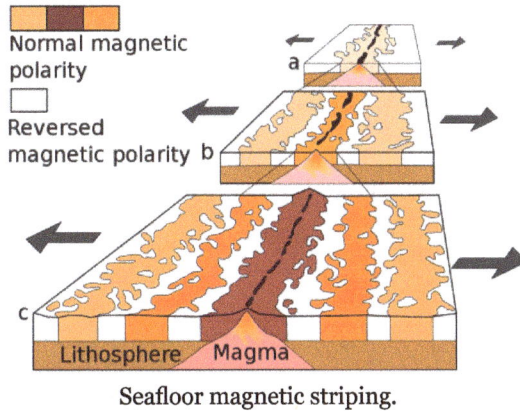

Normal magnetic polarity

Reversed magnetic polarity

Lithosphere   Magma

Seafloor magnetic striping.

A demonstration of magnetic striping. (The darker the color is, the closer it is to normal polarity)

As more and more of the seafloor was mapped during the 1950s, the magnetic variations turned out not to be random or isolated occurrences, but instead revealed recognizable patterns. When these magnetic patterns were mapped over a wide region, the ocean floor showed a zebra-like pattern: one stripe with normal polarity and the adjoining stripe with reversed polarity. The overall pattern, defined by these alternating bands of normally and reversely polarized rock, became known as magnetic striping, and was published by Ron G. Mason and co-workers in 1961, who did not find, though, an explanation for these data in terms of sea floor spreading, like Vine, Matthews and Morley a few years later.

The discovery of magnetic striping called for an explanation. In the early 1960s scientists such as Heezen, Hess and Dietz had begun to theorise that mid-ocean ridges mark structurally weak zones where the ocean floor was being ripped in two lengthwise along the ridge crest. New magma from deep within the Earth rises easily through these weak zones and eventually erupts along the crest of the ridges to create new oceanic crust. This process, at first denominated the "conveyer belt hypothesis" and later called seafloor spreading, operating over many millions of years continues to form new ocean floor all across the 50,000 km-long system of mid-ocean ridges.

Only four years after the maps with the "zebra pattern" of magnetic stripes were published, the link between sea floor spreading and these patterns was correctly placed, independently by Lawrence Morley, and by Fred Vine and Drummond Matthews, in 1963, now called the Vine-Matthews-Morley hypothesis. This hypothesis linked these patterns to geomagnetic reversals and was supported by several lines of evidence:

1. the stripes are symmetrical around the crests of the mid-ocean ridges; at or near the crest of the ridge, the rocks are very young, and they become progressively older away from the ridge crest;

2. the youngest rocks at the ridge crest always have present-day (normal) polarity;

3. stripes of rock parallel to the ridge crest alternate in magnetic polarity (normal-reversed-normal, etc.), suggesting that they were formed during different epochs documenting the (already known from independent studies) normal and reversal episodes of the Earth's magnetic field.

By explaining both the zebra-like magnetic striping and the construction of the mid-ocean ridge system, the seafloor spreading hypothesis (SFS) quickly gained converts and represented another major advance in the development of the plate-tectonics theory. Furthermore, the oceanic crust now came to be appreciated as a natural "tape recording" of the history of the geomagnetic field reversals (GMFR) of the Earth's magnetic field. Today, extensive studies are dedicated to the calibration of the normal-reversal patterns in the oceanic crust on one hand and known timescales derived from the dating of basalt layers in sedimentary sequences (magnetostratigraphy) on the other, to arrive at estimates of past spreading rates and plate reconstructions.

Definition and Refining of the Theory

After all these considerations, Plate Tectonics (or, as it was initially called "New Global Tectonics") became quickly accepted in the scientific world, and numerous papers followed that defined the concepts:

•   In 1965, Tuzo Wilson who had been a promotor of the sea floor spreading hypothesis and continental drift from the very beginning added the concept of transform faults to the model, completing the classes of fault types necessary to make the mobility of the plates on the globe work out.

•   A symposium on continental drift was held at the Royal Society of London in 1965 which must be regarded as the official start of the acceptance of plate tectonics by the scientific community, and which abstracts are issued as Blacket,

Bullard & Runcorn (1965). In this symposium, Edward Bullard and co-workers showed with a computer calculation how the continents along both sides of the Atlantic would best fit to close the ocean, which became known as the famous "Bullard's Fit".

- In 1966 Wilson published the paper that referred to previous plate tectonic reconstructions, introducing the concept of what is now known as the "Wilson Cycle".

- In 1967, at the American Geophysical Union's meeting, W. Jason Morgan proposed that the Earth's surface consists of 12 rigid plates that move relative to each other.

- Two months later, Xavier Le Pichon published a complete model based on 6 major plates with their relative motions, which marked the final acceptance by the scientific community of plate tectonics.

- In the same year, McKenzie and Parker independently presented a model similar to Morgan's using translations and rotations on a sphere to define the plate motions.

Implications for Biogeography

Continental drift theory helps biogeographers to explain the disjunct biogeographic distribution of present-day life found on different continents but having similar ancestors. In particular, it explains the Gondwanan distribution of ratites and the Antarctic flora.

Plate Reconstruction

Reconstruction is used to establish past (and future) plate configurations, helping determine the shape and make-up of ancient supercontinents and providing a basis for paleogeography.

Defining Plate Boundaries

Current plate boundaries are defined by their seismicity. Past plate boundaries within existing plates are identified from a variety of evidence, such as the presence of ophiolites that are indicative of vanished oceans.

Past Plate Motions

Tectonic motion first began around three billion years ago.

Various types of quantitative and semi-quantitative information are available to constrain past plate motions. The geometric fit between continents, such as between west

Africa and South America is still an important part of plate reconstruction. Magnetic stripe patterns provide a reliable guide to relative plate motions going back into the Jurassic period. The tracks of hotspots give absolute reconstructions, but these are only available back to the Cretaceous. Older reconstructions rely mainly on paleomagnetic pole data, although these only constrain the latitude and rotation, but not the longitude. Combining poles of different ages in a particular plate to produce apparent polar wander paths provides a method for comparing the motions of different plates through time. Additional evidence comes from the distribution of certain sedimentary rock types, faunal provinces shown by particular fossil groups, and the position of orogenic belts.

Formation and Break-up of Continents

The movement of plates has caused the formation and break-up of continents over time, including occasional formation of a supercontinent that contains most or all of the continents. The supercontinent Columbia or Nuna formed during a period of 2,000 to 1,800 million years ago and broke up about 1,500 to 1,300 million years ago. The supercontinent Rodinia is thought to have formed about 1 billion years ago and to have embodied most or all of Earth's continents, and broken up into eight continents around 600 million years ago. The eight continents later re-assembled into another supercontinent called Pangaea; Pangaea broke up into Laurasia (which became North America and Eurasia) and Gondwana (which became the remaining continents).

The Himalayas, the world's tallest mountain range, are assumed to have been formed by the collision of two major plates. Before uplift, they were covered by the Tethys Ocean.

Current Plates

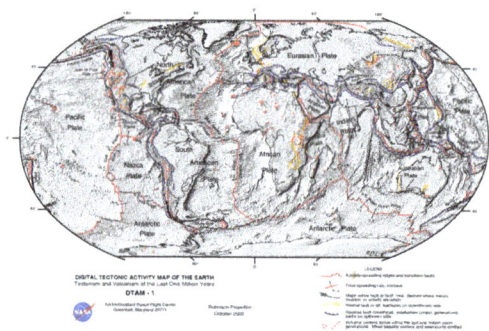

Depending on how they are defined, there are usually seven or eight "major" plates: African, Antarctic, Eurasian, North American, South American, Pacific, and Indo-Australian. The latter is sometimes subdivided into the Indian and Australian plates.

There are dozens of smaller plates, the seven largest of which are the Arabian, Caribbean, Juan de Fuca, Cocos, Nazca, Philippine Sea and Scotia.

The current motion of the tectonic plates is today determined by remote sensing satellite data sets, calibrated with ground station measurements.

Other Celestial Bodies (Planets, Moons)

The appearance of plate tectonics on terrestrial planets is related to planetary mass, with more massive planets than Earth expected to exhibit plate tectonics. Earth may be a borderline case, owing its tectonic activity to abundant water (silica and water form a deep eutectic.)

Venus

Venus shows no evidence of active plate tectonics. There is debatable evidence of active tectonics in the planet's distant past; however, events taking place since then (such as the plausible and generally accepted hypothesis that the Venusian lithosphere has thickened greatly over the course of several hundred million years) has made constraining the course of its geologic record difficult. However, the numerous well-preserved impact craters have been utilized as a dating method to approximately date the Venusian surface (since there are thus far no known samples of Venusian rock to be dated by more reliable methods). Dates derived are dominantly in the range 500 to 750 million years ago, although ages of up to 1,200 million years ago have been calculated. This research has led to the fairly well accepted hypothesis that Venus has undergone an essentially complete volcanic resurfacing at least once in its distant past, with the last event taking place approximately within the range of estimated surface ages. While the mechanism of such an impressive thermal event remains a debated issue in Venusian geosciences, some scientists are advocates of processes involving plate motion to some extent.

One explanation for Venus's lack of plate tectonics is that on Venus temperatures are too high for significant water to be present. The Earth's crust is soaked with water, and water plays an important role in the development of shear zones. Plate tectonics requires weak surfaces in the crust along which crustal slices can move, and it may well be that such weakening never took place on Venus because of the absence of water. However, some researchers remain convinced that plate tectonics is or was once active on this planet.

Mars

Mars is considerably smaller than Earth and Venus, and there is evidence for ice on its surface and in its crust.

In the 1990s, it was proposed that Martian Crustal Dichotomy was created by plate tectonic processes. Scientists today disagree, and think that it was created either by upwelling within the Martian mantle that thickened the crust of the Southern Highlands and formed Tharsis or by a giant impact that excavated the Northern Lowlands.

Valles Marineris may be a tectonic boundary.

Observations made of the magnetic field of Mars by the *Mars Global Surveyor* spacecraft in 1999 showed patterns of magnetic striping discovered on this planet. Some scientists interpreted these as requiring plate tectonic processes, such as seafloor spreading. However, their data fail a "magnetic reversal test", which is used to see if they were formed by flipping polarities of a global magnetic field.

Icy Satellites

Some of the satellites of Jupiter have features that may be related to plate-tectonic style deformation, although the materials and specific mechanisms may be different from plate-tectonic activity on Earth. On 8 September 2014, NASA reported finding evidence of plate tectonics on Europa, a satellite of Jupiter—the first sign of such geological activity on another world other than Earth.

Titan, the largest moon of Saturn, was reported to show tectonic activity in images taken by the *Huygens* probe, which landed on Titan on January 14, 2005.

Exoplanets

On Earth-sized planets, plate tectonics is more likely if there are oceans of water; however, in 2007, two independent teams of researchers came to opposing conclusions about the likelihood of plate tectonics on larger super-earths with one team saying that plate tectonics would be episodic or stagnant and the other team saying that plate tectonics is very likely on super-earths even if the planet is dry.

References

- Dyches, Preston; Brown, Dwayne; Buckley, Michael (8 September 2014). "Scientists Find Evidence of 'Diving' Tectonic Plates on Europa". NASA. Retrieved 8 September 2014.

- Glatzmaier, Gary A. (2013). Introduction to Modeling Convection in Planets and Stars: Magnetic Field, Density Stratification, Rotation. Princeton University Press. p. 149.

- Wolpert, Stuart (August 9, 2012). "UCLA scientist discovers plate tectonics on Mars". Yin, An. UCLA. Retrieved August 13, 2012.

- Neith, Katie (April 15, 2011). "Caltech Researchers Use GPS Data to Model Effects of Tidal Loads on Earth's Surface". Caltech. Retrieved August 15, 2012.

- Kranendonk, V.; Martin, J. (2011). "Onset of Plate Tectonics". Science. 333 (6041): 413–414. Bibcode:2011Sci...333..413V. doi:10.1126/science.1208766. PMID 21778389.

2

Understanding Tectonics

In order to develop a firm understanding of tectonics it is very important to understand concepts such as thrust tectonics, strike-slip tectonics, salt tectonics, neotectonics etc. Thrust tectonics is particularly concerned with the thickening of the crust whereas strike-slip tectonics is concerned with the structures formed by levels of displacement within the crust. This section is an overview of the subject matter incorporating all the major aspects of tectonics.

Tectonics

Tectonics is concerned with the processes which control the structure and properties of the Earth's crust, and its evolution through time. In particular, it describes the processes of mountain building, the growth and behavior of the strong, old cores of continents known as cratons, and the ways in which the relatively rigid plates that constitute the Earth's outer shell interact with each other. Tectonics also provides a framework to understand the earthquake and volcanic belts which directly affect much of the global population. Tectonic studies are important for understanding erosion patterns in geomorphology and as guides for the economic geologist searching for petroleum and metallic ores.

Main Types of Tectonic Regime

Extensional Tectonics

Extensional tectonics is associated with the stretching and thinning of the crust or lithosphere. This type of tectonics is found at divergent plate boundaries, in continental rifts, during and after a period of continental collision caused by the lateral spreading of the thickened crust formed, at releasing bends in strike-slip faults, in back-arc basins and on the continental end of passive margin sequences where a detachment layer is present.

Thrust (Contractional) Tectonics

Thrust tectonics is associated with the shortening and thickening of the crust or lithosphere. This type of tectonics is found at zones of continental collision, at restraining bends in strike-slip faults and at the oceanward part of passive margin sequences where a detachment layer is present.

Strike-slip Tectonics

Strike-slip tectonics is associated with the relative lateral movement of parts of the crust or lithosphere. This type of tectonics is found along oceanic and continental transform faults, at lateral offsets in extensional and thrust fault systems, in the over-riding plate in zones of oblique collision and accommodating deformation in the foreland to a collisional belt.

Plate Tectonics

In plate tectonics the outermost part of the earth – the crust and uppermost mantle – are viewed as acting as a single mechanical layer, the lithosphere. The lithosphere is divided into separate 'plates' that move relative to each other on the underlying, relatively weak asthenosphere in a process ultimately driven by the continuous loss of heat from the earth's interior. There are three main types of plate boundaries: divergent where plates move apart from each other and new lithosphere is formed in the process of sea-floor spreading; transform where plates slide past each other and convergent where plates converge and lithosphere is 'consumed' by the process of subduction. Convergent and transform boundaries form the largest structural discontinuities in the lithosphere and are responsible for most of the world's major ($M_w > 7$) earthquakes. Convergent and divergent boundaries are also the site of most of the world's volcanoes, such as around the Pacific Ring of Fire. Most of the deformation in the lithosphere is related to the interaction between plates, either directly or indirectly.

Other Fields of Tectonic Studies

Salt Tectonics

Salt tectonics is concerned with the structural geometries and deformation processes associated with the presence of significant thicknesses of rock salt within a sequence of rocks. This is due both to the low density of salt, which does not increase with burial, and its low strength.

Neotectonics

Neotectonics is the study of the motions and deformations of the Earth's crust (geological and geomorphological processes) that are current or recent in geological time. The term may also refer to the motions/deformations in question themselves. The corresponding time frame is referred to as the *neotectonic period*. Accordingly, the preceding time is referred to as *palaeotectonic period*.

Tectonophysics

Tectonophysics is the study of the physical processes associated with deformation of the crust and mantle from the scale of individual mineral grains up to that of tectonic plates.

Seismotectonics

Seismotectonics is the study of the relationship between earthquakes, active tectonics and individual faults in a region. It seeks to understand which faults are responsible for seismic activity in an area by analysing a combination of regional tectonics, recent instrumentally recorded events, accounts of historical earthquakes and geomorphological evidence. This information can then be used to quantify the seismic hazard of an area.

Planetary Tectonics

Techniques used in the analysis of tectonics on earth have also been applied to the study of the planets and their moons.

Extensional Tectonics

Extensional tectonics is concerned with the structures formed, and the tectonic processes associated with, the stretching of the crust or lithosphere.

Deformation Styles

The types of structure and the geometries formed depend on the amount of stretching involved. Stretching is generally measured using the parameter β, known as the *beta factor* where

$$\beta = \frac{t_0}{t_1}$$

t_0 is the initial crustal thickness and t_1 is the final crustal thickness. It is also the equivalent of the strain parameter *stretch*.

Low Beta Factor

In areas of relatively low crustal stretching, the dominant structures are high to moderate angle normal faults, with associated half grabens and tilted fault blocks.

High Beta Factor

In areas of high crustal stretching, individual extensional faults may become rotated to too low a dip to remain active and a new set of faults may be generated. Large displacements may juxtapose syntectonic sediments against metamorphic rocks of the mid to lower crust and such structures are called detachment faults. In some cases the detachments are folded such that the metamorphic rocks are exposed within antiformal closures and these are known as metamorphic core complexes.

Passive Margins

Passive margins above a weak layer develop a specific set of extensional structures. Large listric regional (i.e. dipping towards the ocean) faults are developed with rollover anticlines and related crestal collapse grabens. On some margins, such as the Niger Delta, large counter-regional faults are observed, dipping back towards the continent, forming large grabenal mini-basins with antithetic regional faults.

Geological Environments Associated with Extensional Tectonics

Areas of extensional tectonics are typically associated with:

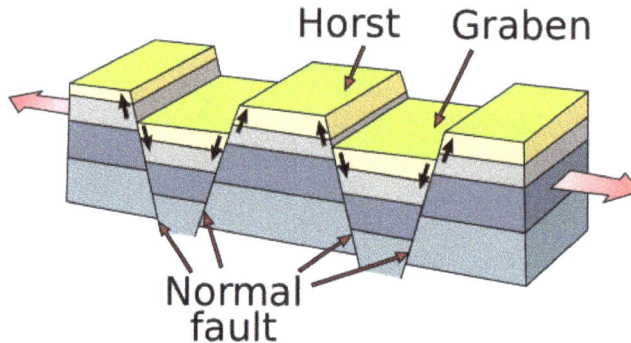

Horst and graben structure, typical rift related structure (direction of extension shown by red arrows).

Continental Rifts

Rifts are linear zones of localized crustal extension. They range in width from somewhat less than 100 km up to several hundred km, consisting of one or more normal faults and related fault blocks. In individual rift segments one polarity (i.e. dip direction) normally dominates giving a half-graben geometry. Other common geometries include metamorphic core complexes and tilted blocks. Examples of active continental rifts are the Baikal Rift Zone and the East African Rift.

Divergent Plate Boundaries

Divergent plate boundaries are zones of active extension as the crust newly formed at the mid-ocean ridge system becomes involved in the opening process.

Gravitational Spreading of Zones of Thickened Crust

Zones of thickened crust, such as those formed during continent-continent collision tend to spread laterally; this spreading occurs even when the collisional event is

still in progress. After the collision has finished the zone of thickened crust generally undergoes gravitational collapse, often with the formation of very large extensional faults. Large-scale Devonian extension, for example, followed immediately
after the end of the Caledonian orogeny particularly in East Greenland and western
Norway.

Releasing Bends Along Strike-slip Faults

When a strike-slip fault is offset along strike such as to create a gap i.e. a left-stepping bend on a sinistral fault, a zone of extension or transtension is generated. Such
bends are known as *releasing bends* or *extensional stepovers* and often form pull-apart
basins or *rhombochasms*. Examples of active pull-apart basins include the Dead Sea,
formed at a left-stepping offset of the sinistral sense Dead Sea Transform system, and
the Sea of Marmara, formed at a right-stepping offset on the dextral sense North Anatolian Fault system.

Back-arc Basins

Back-arc basins form behind many subduction zones due to the effects of oceanic trench
roll-back which leads to a zone of extension parallel to the island arc.

Passive Margins

A passive margin built out over a weaker layer, such as an overpressured mudstone
or salt, tends to spread laterally under its own weight. The inboard part of the sedimentary prism is affected by extensional faulting, balanced by outboard shortening.

Thrust Tectonics

Thrust tectonics or contractional tectonics is concerned with the structures formed,
and the tectonic processes associated with, the shortening and thickening of the crust
or lithosphere.

Deformation Styles

In areas of thrust tectonics two main styles are recognized; thin-skinned deformation and
thick-skinned deformation. The distinction is important as attempts to structurally restore the deformation will give very different results depending on the assumed geometry.

Thin-skinned Deformation

Thin-skinned deformation refers to shortening that only involves the sedimentary cover. This style is typical of many fold and thrust belts developed in the foreland of a collisional zone. This is particularly the case where a good basal decollement exists such as salt or a zone of high pore fluid pressure.

Thick-skinned Deformation

Thick-skinned deformation refers to shortening that involves basement rocks rather than just the overlying cover. This type of geometry is typically found in the hinterland of a collisional zone. This style may also occur in the foreland where no effective decollement surface is present or where pre-existing extensional rift structures may be inverted.

Geological Environments Associated with Thrust Tectonics

Collisional Zones

The most significant areas of thrust tectonics are associated with destructive plate boundaries leading to the formation of orogenic belts. The two main types are: the collision of two continental tectonic plates (for example the Arabian and Eurasian plates, which formed the Zagros fold and thrust belt) and collisions between a continent and an island arc such as that which formed Taiwan.

Restraining Bends on Strike-slip Faults

When a strike-slip fault is offset along strike such that the resulting bend in the fault hinders easy movement, e.g. a right stepping bend on a sinistral (left-lateral) fault, this will cause local shortening or transpression. Examples include the 'Big Bend' region of the San Andreas fault, and parts of the Dead Sea Transform.

Passive Margins

Passive margins are characterised by large prisms of sedimentary material deposited since the original break-up of a continent associated with formation of a new spreading centre. This wedge of material will tend to spread under gravity and, where an effective detachment layer is present such as salt, the extensional faulting that forms at the landward side will be balanced at the front of the wedge by a series of *toe-thrusts*. Examples include the outboard part of the Niger delta (with an overpressured mudstone detachment) and the Angola margin (with a salt detachment).

Strike-slip Tectonics

Strike-slip tectonics is concerned with the structures formed by, and the tectonic processes associated with, zones of lateral displacement within the crust or lithosphere.

Deformation Styles

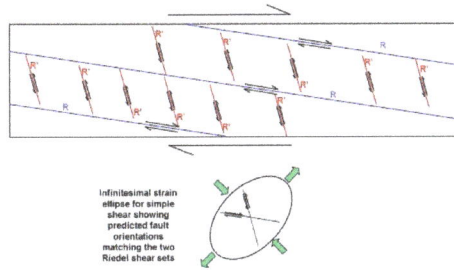

Development of Riedel shears in a zone of dextral shear

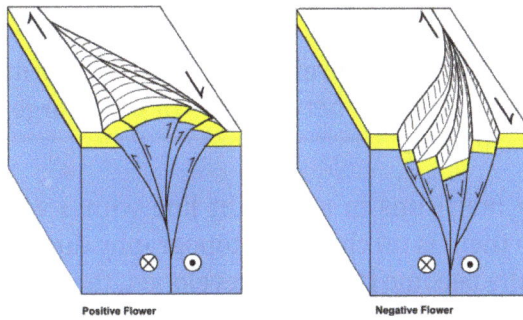

Flower structures developed along minor restraining and releasing bends on a dextral (right-lateral) strike-slip fault

Riedel Shear Structures

In the early stages of strike-slip fault formation, displacement within basement rocks produces characteristic fault structures within the overlying cover. This will also be the case where an active strike-slip zone lies within an area of continuing sedimentation. At low levels of strain the overall simple shear causes a set of small faults to form. The dominant set, known as R shears, form at about 15° to the underlying fault with the same shear sense. The R shears are then linked by a second set, the R' shear that form at about 75° to the main fault trace. These two fault orientations can be understood as conjugate fault sets at 30° to the short axis of the instantaneous strain ellipse associated with the simple shear strain field caused by the displacements applied at the base of the cover sequence. With further displacement the Riedel fault segments will tend to become fully linked, often with the development of a further set of shears known as 'P shears', which are roughly symmetrical to the R shears with respect to the overall shear direction, until a throughgoing fault is formed. The somewhat oblique segments will link downwards into the fault at the base of the cover sequence with a helicoidal geometry.

Flower Structures

In detail many strike-slip faults at surface consist of en echelon and/or braided segments in many cases probably inherited from previously formed Riedel shears. In cross-section the displacements are dominantly reverse or normal in type depending on whether the overall fault geometry is transpressional (i.e. with a small component of shortening) or transtensional (with a small component of extension). As the faults tend to join downwards onto a single strand in basement, the geometry has led to these being termed *flower structure*. Fault zones with dominantly reverse faulting are known as *positive flowers*, those with dominantly normal offsets are known as *negative flowers*. The identification of such structures, particularly where positive and negative flowers are developed on different segments of the same fault, are regarded as reliable indicators of strike-slip.

Strike Slip Duplexes

Strike slip duplexes occur at the step over regions of faults, forming a lens shaped near parallel arrays of horses. These occur between two or more large bounding faults which usually have large displacement.

An idealized strike-slip fault runs in a straight line with a vertical dip and has only horizontal motion, thus there is no change in topography due to motion of the fault. In reality, as strike slip faults become large and developed, their behavior changes and becomes more complex. A long strike slip fault follows a staircase-like trajectory consisting of interspaced fault planes that follow the main fault direction. These sub parallel stretches are isolated by offsets at first, but over long periods of time they can become connected by step overs in order to accommodate the strike slip displacement. In long stretches of strike-slip the fault plane can start to curve, giving rise to structures similar to step overs.

Right lateral motion of a strike slip fault at a right step over (or overstep) gives rise to extensional bends characterised by zones of subsidence, local normal faults, and pull apart basins. On extensional duplexes, normal faults will accommodate the vertical motion, creating negative relief. Similarly, left stepping at a dextral fault generates contractional bends; shortening the step overs which is displayed by local reverse faults, push-up zones, and folds. On contractional duplex structures, thrust faults will accommodate vertical displacement rather than being folded, as the uplifting process is more energy efficient.

Strike slip dulexes are passive structures; they form as a response to displacement of the bounding fault rather than by the stresses from plate motion. Each horse has a length that varies from half to twice the spacing between the bounding fault planes. Depending on the properties of the rocks and the fault, the duplexes will have different length ratios and will develop on either major or subtle offsets, although it is possible to observe duplex structures that develop on nearly straight fault segments. Because the motion of the duplexes may be heterogeneous, the individual horses can experience a rotation with a horizontal axis, which results in the formation of scissor faults. Scissor faults exhibit normal motion at one end of the horse and a thrust motion ant the other end. Because strike slip duplexes structures have more horizontal motion than vertical motion, they are best observed on a map rather than a vertical projection, and are a good indication that the main fault has a strike slip motion.

An example of strike slip duplexes were observed in the Lambertville sill, New Jersey. Flemington and the Hopewell faults, the two main faults in the region, experienced 3 km of dip slip and over 20 km of strike slip motions to accommodate regional extension. It is possible to trace the lensoidal structures which are interpreted as horses that form duplexes. The lens structures observed in the 3M quarry are 180 meters long and 10 meters wide. The main duplex is 30 m in length and other smaller duplexes are also present.

Geological Environments Associated with Strike-slip Tectonics

San Andreas Transform Fault on the Carrizo Plain

Areas of strike-slip tectonics are associated with:

Oceanic Transform Boundaries

Mid-ocean ridges are broken into segments offset from each other by transform faults. The active part of the transform links the two ridge segments. Some of these transforms can be very large, such as the Romanche fracture zone, whose active portion extends for about 300 km.

Continental Transform Boundaries

Transform faults within continental plates include some of the best known examples of strike-slip structures, such as the San Andreas Fault, the Dead Sea Transform, the North Anatolian Fault and the Alpine Fault.

Lateral Ramps in Areas of Extensional or Contractional Tectonics

Major lateral offsets between large extensional or thrust faults are normally connected by diffuse or discrete zones of strike-slip deformation allowing transfer of the overall displacement between the structures.

Zones of Oblique Collision

In most zones of continent-continent collision the relative movement of the plates is oblique to the plate boundary itself. The deformation along the boundary is normally partitioned into dip-slip contractional structures in the foreland with a single large strike-slip structure in the hinterland accommodating all the strike-slip component along the boundary. Examples include the *Main Recent Fault* along the boundary between the Arabian and Eurasian plates behind the Zagros fold and thrust belt, the Liquiñe-Ofqui Fault that runs through Chile and the Great Sumatran fault that runs parallel to the subduction zone along the Sunda Trench.

The Deforming Foreland of a Zone of Continent-continent Collision

The process sometimes known as indenter tectonics, first elucidated by Paul Tapponnier, occurs during a collisional event where one of the plates deforms internally along a system of strike-slip faults. The best known active example is the system of strike-slip structures observed in the Eurasian plate as it responds to collision with the Indian plate, such as the Kunlun fault and Altyn Tagh fault.

Salt Tectonics

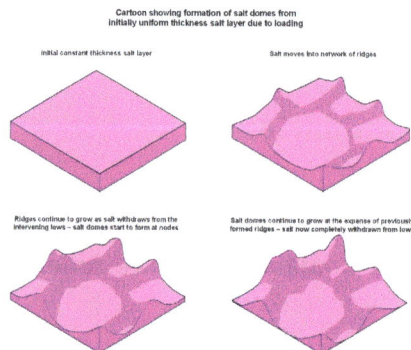

Cartoon showing formation of salt domes from initially uniform thickness salt layer due to loading

Salt tectonics is concerned with the geometries and processes associated with the presence of significant thicknesses of evaporites containing rock salt within a stratigraphic sequence of rocks. This is due both to the low density of salt, which does not increase with burial, and its low strength.

Passive salt structures

Structures may form during continued sedimentary loading, without any external tectonic influence, due to gravitational instability. Pure halite has a density of 2160 kg/m³. When initially deposited, sediments generally have a lower density of 2000 kg/m³, but with loading and compaction their density increases to 2500 kg/m³, which is greater than that of salt. Once the overlying layers have become denser, the weak salt layer will tend to deform into a characteristic series of ridges and depressions, due to a form of Rayleigh–Taylor instability. Further sedimentation will be concentrated in the depressions and the salt will continue to move away from them into the ridges. At a late stage, diapirs tend to initiate at the junctions between ridges, their growth fed by movement of salt along the ridge system, continuing until the salt supply is exhausted. During the later stages of this process the top of the diapir remains at or near the surface, with further burial being matched by diapir rise, and is sometimes referred to as *downbuilding*. The Schacht Asse II and Gorleben salt domes in Germany are an example of a purely passive salt structure.

Such structures do not always form when a salt layer is buried beneath a sedimentary overburden. This can be due to a relatively high strength overburden or to the presence of sedimentary layers interbedded within the salt unit that increase both its density and strength.

Active Salt Structures

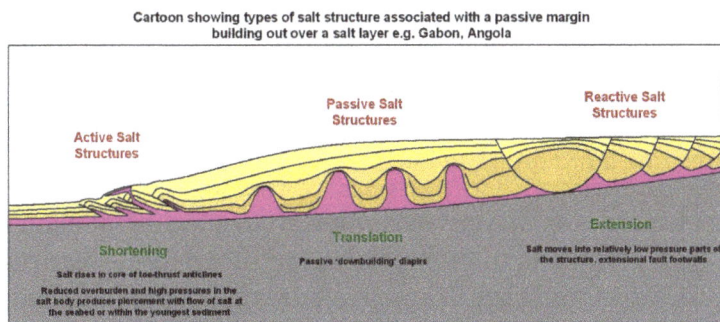

Cartoon showing types of salt structure associated with a passive margin building out over a salt layer e.g. Gabon, Angola

Active tectonics will increase the likelihood of salt structures developing. In the case of extensional tectonics, faulting will both reduce the strength of the overburden and thin it. In an area affected by thrust tectonics, buckling of the overburden layer will allow the salt to rise into the cores of anticlines, as seen in salt domes in the Zagros Mountains.

If the pressure within the salt body becomes sufficiently high it may be able to push through its overburden, this is known as *forceful* diapirism. Many salt diapirs may contain elements of both active and passive salt movement. An active salt structure may pierce its overburden and from then on continue to develop as a purely passive salt diapir.

Reactive salt structures

In those cases where salt layers do not have the conditions necessary to develop passive salt structures, the salt may still move into relatively low pressure areas around developing folds and faults. Such structures are described as *reactive*.

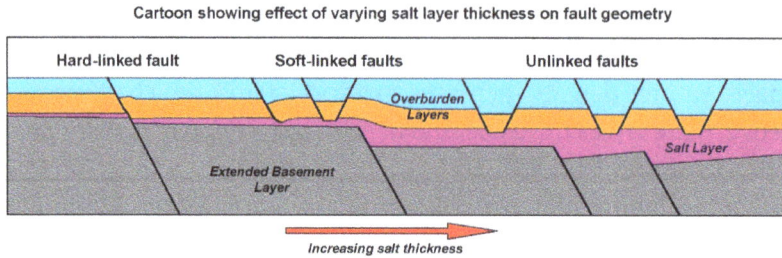

Cartoon showing effect of varying salt layer thickness on fault geometry

Salt Detached Fault Systems

When one or more salt layers are present during extensional tectonics, a characteristic set of structures are formed. Extensional faults propagate up from the middle part of the crust until they encounter the salt layer. The weakness of the salt prevents the fault from propagating through. However, continuing displacement on the fault offsets the base of the salt and causes bending of the overburden layer. Eventually the stresses caused by this bending will be sufficient to fault the overburden. The types of structures developed depend on the initial salt thickness. In the case of a very thick salt layer there is no direct spatial relationship between the faulting beneath the salt and that in the overburden, such a system is said to be *unlinked*. For intermediate salt thicknesses, the overburden faults are spatially related to the deeper faults, but offset from them, normally into the footwall; these are known as *soft-linked* systems. When the salt layer becomes thin enough, the fault that develops in the overburden is closely aligned with that beneath the salt, and forms a continuous fault surface after only a relatively small displacement, forming a *hard-linked* fault.

In areas of thrust tectonics salt layers act as preferred detachment planes. In the Zagros fold and thrust belt, variations in the thickness and therefore effectiveness of the late Neoproterozoic to Early Cambrian Hormuz salt are thought to have had a fundamental control on the overall topography.

Salt Weld

When a salt layer becomes too thin to be an effective detachment layer, due to salt movement, dissolution or removal by faulting, the overburden and the underlying sub-salt basement become effectively *welded* together. This may cause the development of new faults in the cover sequence and is an important consideration when modeling the migration of hydrocarbons. Salt welds may also develop in the vertical direction by putting the sides of a former diapir in contact.

Allochthonous Salt Structures

Salt that pierces to the surface, either on land or beneath the sea, tends to spread laterally away and such salt is said to be "allochthonous". Salt glaciers are formed on land where this happens in an arid environment, such as in the Zagros Mountains. Offshore tongues of salt are generated that may join together with others from neighbouring piercements to from canopies.

Economic Importance

A significant proportion of the world's hydrocarbon reserves are found in structures related to salt tectonics, including many in the Middle East, the South Atlantic passive margins (Brazil, Gabon and Angola) and the Gulf of Mexico.

Neotectonics

Neotectonics, a subdiscipline of tectonics, is the study of the motions and deformations of Earth's crust (geological and geomorphological processes) that are current or recent in geologic time. The term may also refer to the motions/deformations in question themselves. Geologists refer to the corresponding time-frame as the neotectonic period, and to the preceding time as the palaeotectonic period.

Vladimir Obruchev coined the term *neotectonics* in his 1948 article, defining the field as "recent tectonic movements occurred in the upper part of Tertiary (Neogene) and in the Quaternary, which played an essential role in the origin of the contemporary topography". Since then geologists have disagreed as to how far back to date "geologically recent" time, with the common meaning being that neotectonics is the youngest, not yet finished stage in Earth tectonics. Some authors consider neotectonics to be basically synonymous with "active tectonics", while others date the start of the neotectonic period from the middle Miocene. A general agreement has started to emerge that the actual time-frame may be individual for each geological environment and it must be set back in time sufficiently far to fully understand the current tectonic activity.

In 1989 Spyros B. Pavlides suggested the definition:

"Neotectonics is the study of young tectonic events which have occurred or are still occurring in a given region after its orogeny or after its last significant tectonic set-up [...] The tectonic events are recent enough to permit a detailed analysis by differentiated and specific methods, while their results are directly compatible with seismological observations."

Many researchers have accepted this approach.

The Center for Neotectonic Studies at the University of Nevada, Reno defines neotectonics as

"the study of geologically recent motions of the Earth's crust, particularly those produced by earthquakes, with the goals of understanding the physics of earthquake recurrence, the growth of mountains, and the seismic hazard embodied in these processes."

One source of different interpretations for a region stems from the fact that changes in different tectonic plates of the region may occur at different times, giving rise to the notion of the "transitional time", during which both palaeotectonic and neotectonic features coexist. For example, for central/northern Europe, the transitional period stretches from the middle early Miocene to the Miocene-Pliocene boundary.

References

- Edward A. Keller (2001) Active Tectonics: Earthquakes, Uplift, and Landscape Prentice Hall; 2nd edition, ISBN 0-13-088230-5

- Stanley A. Schumm, Jean F. Dumont and John M. Holbrook (2002) Active Tectonics and Alluvial Rivers, Cambridge University Press; Reprint edition, ISBN 0-521-89058-6

- B.A. van der Pluijm and S. Marshak (2004). Earth Structure - An Introduction to Structural Geology and Tectonics. 2nd edition. New York: W.W. Norton. p. 656. ISBN 0-393-92467-X.

- McGeary. D and C. C. Plummer (1994) Physical Geology: Earth revealed, Wm . C. Brown Publishers, Dubuque, p.475-476 ISBN 0-697-12687-0

- "Encyclopedia of Coastal Science" (2005), Springer, ISBN 978-1-4020-1903-6, Chapter 1: "Tectonics and Neotectonics" doi:10.1007/1-4020-3880-1

3

Major Tectonic Plates

Pacific Plate is the tectonic plate that exists underneath the Pacific Ocean. The North American Plate similarly covers North America, Greenland and Cuba. The other major tectonic plates covered within this section are Eurasian Plate, African Plate, Antarctic Plate, Australian Plate, South American Plate, Scotia Plate and Indian plate. This chapter elucidates the main tectonic plates.

Pacific Plate

The Pacific Plate is an oceanic tectonic plate that lies beneath the Pacific Ocean. At 103 million square kilometres (40,000,000 sq mi), it is the largest tectonic plate.

The Pacific Plate contains an interior hot spot forming the Hawaiian Islands.

Hillis and Müller are reported to consider the Bird's Head Plate to be moving in unison with the Pacific Plate. Bird considers them to be unconnected.

Boundaries

The north-eastern side is a divergent boundary with the Explorer Plate, the Juan de Fuca Plate and the Gorda Plate forming respectively the Explorer Ridge, the Juan de Fuca Ridge and the Gorda Ridge. In the middle of the eastern side is a transform boundary with the North American Plate along the San Andreas Fault, and a boundary with the Cocos Plate. The south-eastern side is a divergent boundary with the Nazca Plate forming the East Pacific Rise.

The southern side is a divergent boundary with the Antarctic Plate forming the Pacific-Antarctic Ridge.

The western side, the plate is bounded by the Okhotsk Plate at the Kuril-Kamchatka Trench and the Japan Trench, forms a convergent boundary by subducting under the Philippine Sea Plate creating the Mariana Trench, has a transform boundary with the Caroline Plate, and has a collision boundary with the North Bismarck Plate.

In the south-west, the Pacific Plate has a complex but generally convergent boundary with the Indo-Australian Plate, subducting under it north of New Zealand forming the Tonga Trench and the Kermadec Trench. The Alpine Fault marks a transform boundary between the two plates, and further south the Indo-Australian Plate subducts under the Pacific Plate forming the Puysegur Trench. The southern part of Zealandia, which is to the east of this boundary, is the plate's largest block of continental crust.

The northern side is a convergent boundary subducting under the North American Plate forming the Aleutian Trench and the corresponding Aleutian Islands.

Paleo-geology of the Pacific Plate

The Pacific Plate is almost entirely oceanic crust, but it contains some continental crust in New Zealand, Baja California, and coastal California.

The Pacific Plate has the distinction of showing one of the largest areal sections of the oldest members of seabed geology being entrenched into eastern Asian oceanic trenches. A geologic map of the Pacific Ocean seabed shows not only the geologic sequences, and associated Ring of Fire zones on the ocean's perimeters, but the various ages of the seafloor in a stairstep fashion, youngest to oldest, the oldest being consumed into the Asian oceanic trenches. The oldest member disappearing by way of the Plate Tectonics cycle is early-Cretaceous (145 to 137 million years ago).

All maps of the Earth's ocean floor geology show ages younger than 145 million years, only about 1/30 of the Earth's 4.55 billion year history.

North American Plate

The North American Plate is a tectonic plate covering most of North America, Greenland, Cuba, the Bahamas, extreme northeastern Russia, and parts of Iceland and the Azores. It extends eastward to the Mid-Atlantic Ridge and westward to the Chersky Range in eastern Siberia. The plate includes both continental and oceanic crust. The interior of the main continental landmass includes an extensive granitic core called a craton. Along most of the edges of this craton are fragments of crustal material called terranes, accreted to the craton by tectonic actions over the long span of geologic time. It is thought that much of North America west of the Rocky Mountains is composed of such terranes.

Boundaries

The southerly boundary with the Cocos Plate to the west and the Caribbean Plate to the east is a transform fault, represented by the Cayman Trench under the Caribbean Sea and the Motagua Fault through Guatemala. The parallel Septentrional and Enriquillo-Plantain Garden faults, which run through the island of Hispaniola and bound the Gonâve Microplate, are also a part of the boundary. The rest of the southerly margin which extends east to the Mid Atlantic Ridge and marks the boundary between the

North American Plate and the South American Plate is vague but located near the Fifteen-Twenty Fracture Zone around 16°N.

On the northerly boundary is a continuation of the Mid-Atlantic ridge called the Gakkel Ridge. The rest of the boundary in the far northwestern part of the plate extends into Siberia. This boundary continues from the end of the Gakkel Ridge as the Laptev Sea Rift, on to a transitional deformation zone in the Chersky Range, then the Ulakhan Fault between it and the Okhotsk Plate, and finally the Aleutian Trench to the end of the Queen Charlotte Fault system.

The westerly boundary is the Queen Charlotte Fault running offshore along the coast of Alaska and the Cascadia subduction zone to the north, the San Andreas Fault through California, the East Pacific Rise in the Gulf of California, and the Middle America Trench to the south.

On its western edge the Farallon Plate has been subducting under the North American Plate since the Jurassic Period. The Farallon Plate has almost completely subducted beneath the western portion of the North American Plate leaving that part of the North American Plate in contact with the Pacific Plate as the San Andreas Fault. The Juan de Fuca, Explorer, Gorda, Cocos and Nazca Plates are remnants of the Farallon Plate.

The boundary along the Gulf of California is complex. The Gulf is underlain by the Gulf of California Rift Zone, a series of rift basins and transform fault segments between the northern end of the East Pacific Rise in the mouth of the gulf to the San Andreas Fault system in the vicinity of the Salton Trough rift/Brawley seismic zone.

It is generally accepted that a piece of the North American Plate was broken off and transported north as the East Pacific Rise propagated northward, creating the Gulf of California. However, it is as yet unclear whether the oceanic crust east of the Rise and west of the mainland coast of Mexico is actually a new plate beginning to converge with the North American Plate, consistent with the standard model of rift zone spreading centers generally.

Hotspots

A few hotspots are thought to exist below the North American Plate. The most notable hotspots are the Yellowstone (Wyoming), Raton (New Mexico), and Anahim (British Columbia) hotspots. These are thought to be caused by a narrow stream of hot mantle convecting up from the Earth's core-mantle boundary called a mantle plume, although some geologists prefer upper-mantle convection as a cause. The Yellowstone and Anahim hotspots are thought to have first arrived during the Miocene period and are still geologically active, creating earthquakes and volcanoes. The Yellowstone hotspot is most notable for the Yellowstone Caldera and the many calderas that lie in the Snake

River Plain while the Anahim hotspot is most notable for the Anahim Volcanic Belt, currently found in the Nazko Cone area.

Plate Motion

For the most part, the North American Plate moves in roughly a southwest direction away from the Mid-Atlantic Ridge.

The motion of the plate cannot be driven by subduction as no part of the North American Plate is being subducted, except for a small section comprising part of the Puerto Rico Trench; thus other mechanisms continue to be investigated.

One recent study suggests that a mantle convective current is propelling the plate.

Eurasian Plate

The Eurasian Plate is a tectonic plate which includes most of the continent of Eurasia (a landmass consisting of the traditional continents of Europe and Asia), with the notable exceptions of the Indian subcontinent, the Arabian subcontinent, and the area east of the Chersky Range in East Siberia. It also includes oceanic crust extending westward to the Mid-Atlantic Ridge and northward to the Gakkel Ridge.

The eastern side is a boundary with the North American Plate to the north and a boundary with the Philippine Sea Plate to the south, and possibly with the Okhotsk Plate and the Amurian Plate. The southerly side is a boundary with the African Plate to the west, the Arabian Plate in the middle and the Indo-Australian Plate to the east. The westerly side is a divergent boundary with the North American Plate forming the northernmost part of the Mid-Atlantic Ridge, which is straddled by Iceland. All of the volcanic eruptions in Iceland, such as the 1973 eruption of Eldfell, the 1783 eruption of Laki, and the 2010 eruption of Eyjafjallajökull, are caused by the North American and the Eurasian plates moving apart, which is a result of divergent plate boundary forces.

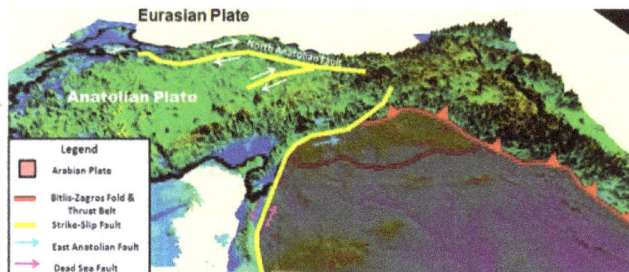

Eurasian & Anatolian Plates

The geodynamics of central Asia is dominated by the interaction between the Eurasian and Indian Plates. In this area, many subplates or crust blocks have been recognized, which form the Central Asian and the East Asian transit zones.

African Plate

The African Plate is a major tectonic plate straddling the equator as well as the prime meridian. It includes much of the continent of Africa, as well as oceanic crust which lies between the continent and various surrounding ocean ridges. Between 60 million years ago and 10 million years ago, the Somali Plate began rifting from the African Plate along the East African Rift. Since the continent of Africa consists of crust from both the African and the Somali plates, some literature refers to the African Plate as the Nubian Plate to distinguish it from the continent as a whole.

Boundaries

The western edge of the African Plate is a divergent boundary with the North American Plate to the north and the South American Plate to the south which forms the central and southern part of the Mid-Atlantic Ridge. The African plate is bounded on the northeast by the Arabian Plate, the southeast by the Somali Plate, the north by the Eurasian Plate, the Aegean Sea Plate, and the Anatolian Plate, and on the south by the Antarctic Plate. All of these are divergent or spreading boundaries with the exception of the northern boundary and a short segment near the Azores known as the Terceira Rift.

Components

The African Plate includes several cratons, stable blocks of old crust with deep roots in the subcontinental lithospheric mantle, and less stable terranes, which came together to form the African continent during the assembly of the supercontinent Pangea around 550 million years ago. The cratons are, from south to north, the Kalahari craton, Congo craton, Tanzania craton and West African craton. The cratons were widely separated in the past, but came together during the Pan-African orogeny and stayed together when Gondwana split up. The cratons are connected by orogenic belts, regions of highly deformed rock where the tectonic plates have engaged. The Saharan Metacraton has been tentatively identified as the remains of a craton that has become detached from the subcontinental lithospheric mantle, but alternatively may consist of a collection of unrelated crustal fragments swept together during the Pan-African orogeny.

In some areas, the cratons are covered by sedimentary basins, such as the Tindouf basin, Taoudeni basin and Congo basin, where the underlying archaic crust is overlaid by more recent Neoproterozoic sediments. The plate includes shear zones such as the Central African Shear Zone (CASZ) where, in the past, two sections of the crust were moving in opposite directions, and rifts such as the Anza trough where the crust was pulled apart, and the resulting depression filled with more modern sediment.

Modern Movements

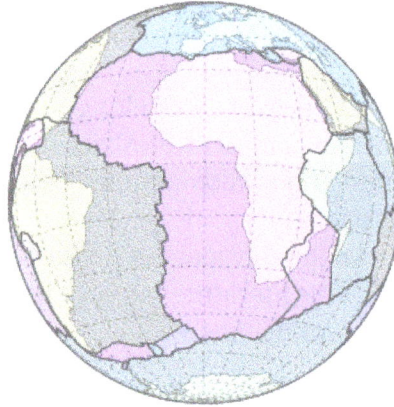

Today, African Plate is moving around the surface of the earth at the speed of 0.292° +- 0.007° per million years, relative to the "average" Earth (NNR-MORVEL56)

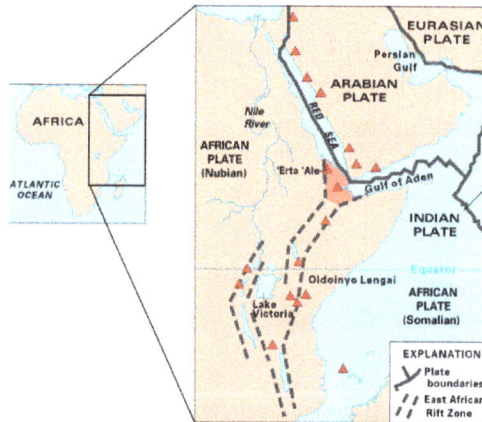

Map of East Africa showing some of the historically active volcanoes(red triangles) and the Afar Triangle (shaded, center) -- a triple junction where three plates are pulling away from one another: the Arabian Plate, the African Plate, and the Somali Plate (USGS).

The African Plate is rifting in the eastern interior of the African continent along the East African Rift. This rift zone separates the African Plate to the west from the Somali Plate to the east. One hypothesis proposes the existence of a mantle plume beneath the Afar region, whereas an opposing hypothesis asserts that the rifting is merely a zone of maximum weakness where the African Plate is deforming as plates to its east are moving rapidly northward.

The African Plate's speed is estimated at around 2.15 cm (0.85 in) per year. It has been moving over the past 100 million years or so in a general northeast direction. This is drawing it closer to the Eurasian Plate, causing subduction where oceanic crust is converging with continental crust (e.g. portions of the central and eastern Mediterranean). In the western Mediterranean, the relative motions of the Eurasian and African plates produce a combination of lateral and compressive forces, concentrated in a zone known as the Azores–Gibraltar Fault Zone. Along its northeast margin, the African Plate is bounded by the Red Sea Rift where the Arabian Plate is moving away from the African Plate.

The New England hotspot in the Atlantic Ocean has probably created a short line of mid- to late-Tertiary age seamounts on the African Plate but appears to be currently inactive.

Antarctic Plate

The Antarctic Plate is a tectonic plate containing the continent of Antarctica and extending outward under the surrounding oceans. After breakup from Gondwana (the southern part of the supercontinent Pangea), the Antarctic plate began moving the continent of Antarctica south to its present isolated location causing the continent to develop a much colder climate. The Antarctic Plate is bounded almost entirely by extensional mid-ocean ridge systems. The adjoining plates are the Nazca Plate, the South American Plate, the African Plate, the Indo-Australian Plate, the Pacific Plate, and, across a transform boundary, the Scotia Plate.

The Antarctic plate has an area of about 60,900,000 km² (23,500,000 sq mi). It is the Earth's fifth largest plate.

The Antarctic plate's movement is estimated to be at least 1 cm (0.4 in) per year towards the Atlantic Ocean.

Australian Plate

The Australian Plate is a major tectonic plate in the eastern and, largely, in the southern hemispheres. Originally a part of the ancient continent of Gondwana, Australia remained connected to India and Antarctica until approximately 100 million years ago when India broke away and began moving north. Australia and Antarctica began rifting 85 million years ago and completely separated roughly 45 million years ago The Australian plate later fused with the adjacent Indian Plate beneath the Indian Ocean to form a single Indo-Australian Plate. However, recent studies suggest that the two plates have once again split apart and have been separate plates for at least 3 million years and likely longer. The Australian plate includes the continent of Australia, including Tasmania, as well portions of New Guinea, New Zealand, and the Indian Ocean basin.

Geography

The northeasterly side is a complex but generally convergent boundary with the Pacific Plate. The Pacific Plate is subducting under the Australian Plate, which forms the Tonga and Kermadec Trenches, and the parallel Tonga and Kermadec island arcs. It has also uplifted the eastern parts of New Zealand's North Island.

The continent of Zealandia, which separated from Australia 85 million years ago and stretches from New Caledonia in the north to New Zealand's subantarctic islands in the south, is now being torn apart along the transform boundary marked by the Alpine Fault.

South of New Zealand the boundary becomes a transitional transform-convergent boundary, the Macquarie Fault Zone, where the Australian Plate is beginning to subduct under the Pacific Plate along the Puysegur Trench. Extending southwest of this trench is the Macquarie Ridge.

The southerly side is a divergent boundary with the Antarctic Plate called the Southeast Indian Ridge (SEIR).

The subducting boundary through Indonesia is not parallel to the biogeographical Wallace line that separates the indigenous fauna of Asia from that of Australasia. The eastern islands of Indonesia lie mainly on the Eurasian Plate, but have Australasian-related fauna and flora. Southeasterly lies the Sunda Shelf.

Origins

Depositional age of the Mount Barren Group on the southern margin of the Yilgarn Craton and zircon provenance analysis support the hypothesis that collisions between the Pilbara–Yilgarn and Yilgarn–Gawler Cratons assembled a proto-Australian continent approximately 1,696 million years ago (Dawson et al. 2002).

Indian Plate

The Indian Plate or India Plate is a major tectonic plate straddling the equator in the eastern hemisphere. Originally a part of the ancient continent of Gondwana, India broke away from the other fragments of Gondwana 100 million years ago and began moving north. Once fused with the adjacent Australia to form a single Indo-Australian Plate, recent studies suggest that India and Australia have been separate plates for at least 3 million years and likely longer. The Indian plate includes most of South Asia—i.e. the Indian subcontinent—and a portion of the basin under the Indian Ocean, including parts of South China and Eastern Indonesia, and extending up to but not including Ladakh, Kohistan and Balochistan.

Plate Movements

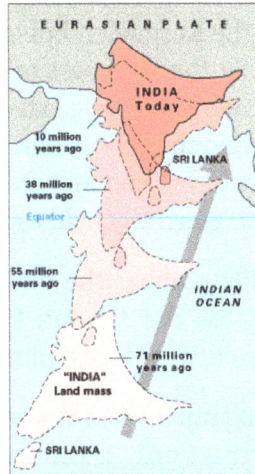

Due to plate tectonics, the India Plate split from Madagascar and collided (c. 55 Ma) with the Eurasian Plate, resulting in the formation of the Himalayas.

Until roughly 140 million years ago, the Indian Plate formed part of the supercontinent Gondwana together with modern Africa, Australia, Antarctica, and South America. Gondwana broke up as these continents drifted apart at different velocities, a process which led to the opening of the Indian Ocean.

In the late Cretaceous, approximately 100 million years ago and subsequent to the splitting off from Gondwana of conjoined Madagascar and India, the Indian Plate split from Madagascar. It began moving north, at about 20 centimetres (7.9 in) per year, and is believed to have begun colliding with Asia as early as 55 million years ago, in the Eocene epoch of the Cenozoic. However, some authors suggest that the collision between India and Eurasia occurred much later, around 35 million years ago. If the collision occurred between 55 and 50 Mya, the Indian Plate would have covered a distance of 3,000 to 2,000 kilometres (1,900 to 1,200 mi), moving faster than any other known plate. In 2012, paleomagnetic data from the Greater Himalaya was used to propose two collisions to reconcile the discrepancy between the amount of crustal shortening in the Himalaya (~1300 km) and the amount of convergence between India and Asia (~3600 km). These authors propose a continental fragment of northern Gondwana rifted from India, traveled northward, and initiated the "soft collision" between the Greater Himalaya and Asia at ~50 Ma. This was followed by the "hard collision" between India and Asia occurred at ~25 Ma. Subduction of the resulting ocean basin that formed between the Greater Himalayan fragment and India explains the apparent discrepancy between the crustal shortening estimates in the Himalaya and paleomagnetic data from India and Asia.

In 2007, German geologists suggested that the reason the Indian Plate moved so quickly is that it is only half as thick (100 kilometres or 62 miles) as the other plates which formerly constituted Gondwana. The mantle plume that once broke up Gondwana might also have melted the lower part of the Indian subcontinent, which allowed it to move both faster and further than the other parts. The remains of this plume today form the Marion Hotspot (Prince Edward Islands), the Kerguelen hotspot, and the Réunion hotspots. As India moved north, it is possible that the thickness of the Indian plate degenerated further as it passed over the hotspots and magmatic extrusions associated with the Deccan and Rajmahal Traps. The massive amounts of volcanic gases released during the passage of the Indian Plate over the hotspots have been theorised to have played a role in the Cretaceous–Paleogene extinction event, generally held to be due to a large asteroid impact.

The collision with the Eurasian Plate along the boundary between India and Nepal formed the orogenic belt that created the Tibetan Plateau and the Himalaya Mountains, as sediment bunched up like earth before a plow.

The Indian Plate is currently moving north-east at 5 centimetres (2.0 in) per year, while the Eurasian Plate is moving north at only 2 centimetres (0.79 in) per year. This is causing the Eurasian Plate to deform, and the Indian Plate to compress at a rate of 4 millimetres (0.16 in) per year.

Geography

The westerly side of the Indian Plate is a transform boundary with the Arabian Plate called the Owen Fracture Zone, and a divergent boundary with the African Plate called the Central Indian Ridge (CIR). The northerly side of the Plate is a convergent boundary with the Eurasian Plate forming the Himalaya and Hindu Kush mountains.

South American Plate

The South American Plate (Dutch: *Zuid-Amerikaanse Plaat*, French: *Plaque Sud-américaine*, Portuguese: *Placa Sul-Americana*, Spanish: *Placa Sudamericana*) is a tectonic plate which includes the continent of South America and also a sizeable region of the Atlantic Ocean seabed extending eastward to the Mid-Atlantic Ridge.

The easterly side is a divergent boundary with the African Plate forming the southern part of the Mid-Atlantic Ridge. The southerly side is a complex boundary with the Antarctic Plate and the Scotia Plate. The westerly side is a convergent boundary with the subducting Nazca Plate. The northerly side is a boundary with the Caribbean Plate and the oceanic crust of the North American Plate. At the Chile Triple Junction in Taitato-Tres Montes Peninsula, an oceanic ridge — the Chile Rise — is subducting under the South American plate.

The South American Plate is in motion. "Parts of the plate boundaries consisting of alternations of relatively short transform fault and spreading ridge segments are represented by a boundary following the general trend." Moving westward away from the Mid-Atlantic Ridge. The eastward-moving and more dense Nazca Plate is subducting under the western edge of the South American Plate along the Pacific coast of the continent at a rate of 77 mm (3.0 in) per year. This collision of plates is responsible for lifting the massive Andes Mountains and causing the volcanoes which are strewn throughout them.

Scotia Plate

The Scotia Plate (Spanish: *Placa Scotia*) is a tectonic plate on the edge of the South Atlantic and Southern Ocean. Thought to have formed during the early Eocene with the opening of the Drake Passage that separates South America from Antarctica, it is a minor plate whose movement is largely controlled by the two major plates that surround it: the South American plate and Antarctic plate.

Roughly rhomboid, extending between 50°S 70°W50°S 70°W and 63°S 20°W63°S 20°W, the plate is 800 km (500 mi) wide and 3,000 km (1,900 mi) long. It is moving WSW at 2.2 cm (0.87 in)/year and the South Sandwich Plate is moving east at 5.5 cm (2.2 in)/year in an absolute reference frame. It takes it name from the steam yacht *Scotia* of the Scottish National Antarctic Expedition (1902–04), the expedition that made the first bathymetric study of the region.

The Scotia Plate is made of oceanic crust and continental fragments now distributed around the Scotia Sea. Before the formation of the plate began 40 million years ago (Ma), these fragments formed a continuous landmass from Patagonia to the Antarctic Peninsula along an active subduction margin. At present the plate is almost completely submerged, with only the small exceptions of the South Georgia Islands on its north-eastern edge and the southern tip of South America.

Tectonic Setting

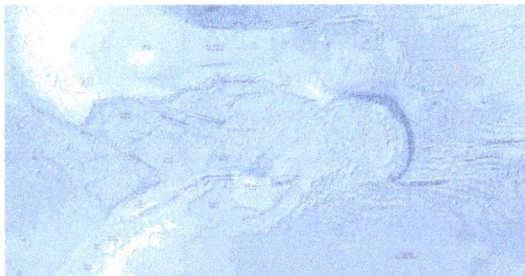

Bathymetric map of Scotia Plate

Together with the Sandwich Plate, the Scotia Plate joins the southern-most Andes to the Antarctic Peninsula, just like the Caribbean Plate joins the north-most Andes to North America, and these two plates are comparable in several ways. Both have volcanic arcs at their eastern ends, the South Sandwich Islands on the Scotia Plate and the Lesser Antilles on the Caribbean Plate, and both plates also had a major impact on global climate when they closed the two major gateways between the Pacific and Atlantic Oceans during the Mesozoic and Cenozoic.

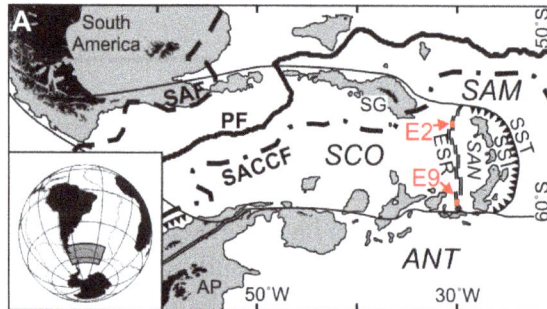

The Scotia plate (SCO) bounded by the South Sandwich plate (SAN), Antarctic plate (ANT), South American plate (SAM) and the Shetland plate, seen just above the Antarctic Peninsula (AP).

North Scotia Ridge

The northern edge of the Scotia plate is bounded by the South American plate, forming the North Scotia Ridge. The North Scotia ridge is a left-lateral transform boundary with a transform rate of roughly 7.1 mm/yr. The Magallanes–Fagnano Fault is passing through Tierra del Fuego.

The northern ridge stretches from Isla de los Estados off Tierra del Fuego in the west to the microcontinent South Georgia in the east, with a series of shallow banks in between: Burdwood, Davis, Barker, and Shag Rocks. North of the ridge is the 3 km (1.9 mi) deep Falkland Trough.

South Georgia Microcontinent

Experts in plate tectonics have been unable to determine whether the South Georgian Islands are part of the Scotia plate or have been recently accreted to the South American plate. Surface expressions of the plate boundary are found north of the islands suggesting a long-term presence of the transform fault there. Yet seismic studies have identified strain and thrusting south of the islands indicating the possible shift of the transform fault to an area south of the island. It has also been suggested that the plate bearing the islands may have broken off from the Scotia plate, forming a new independent South Georgia microplate, yet there is little evidence to make this conclusion.

The South Georgia microcontinent was originally connected to the Roca Verdes back-arc basin (southern-most Tierra del Fuego) until the Eocene. Before that, this basin went through a series of geological transformations during the Cretaceous, through which South Georgia was first buried, then made a topographic feature again by the Late Cretaceous. At about 45 Ma, South Georgia, still part of the South American Plate, got buried again and something, possibly rotation of the Fuegian Andes, completed the break-up and allowed South Georgia a second exhumation. During the Oligocene (34-

23 Ma) South Georgia was reburied again as seafloor spreading took place in the West Scotia Sea. 10 Ma, finally, the South Georgia microcontinent was uplifted as a result of the collision with the Northeast Georgia Rise.

South Scotia Ridge

The southern edge of the plate is bordered by the Antarctic plate, forming the South Scotia Ridge, a left-lateral transform boundary sliding at a rate of roughly 7.4-9.5 mm/yr that occupies the southern half of the Antarctic-Scotia plate boundary. The relative motion between the Scotia plate and the Antarctic plate on the western boundary is 7.5-8.7 mm/yr. Though the South Scotia Ridge is overall a transform fault, small sections of the ridge are spreading to make up for the somewhat jagged shape of the boundary.

At the eastern tip of the Antarctic Peninsula, the beginning of South Scotia Ridge, a small and heavily-dissected bank degenerates into several outcrops dominated by Paleozoic and Cretaceous rocks. A small basin, Powel Basin, separates this cluster from the South Orkney microcontinent composed of Triassic and younger rocks.

The eastern continuation of the ridge, the Scotia Arc east of the South Sandwich Plate, are the South Sandwich island arc and trench. This volcanic active island arc has submerged ancestors in Jane and Discovery banks in the southern ridge.

Shackleton Fracture Zone

The western edge of the plate is bounded by the Antarctic plate, forming the Shackleton Fracture Zone and the southern Chile Trench. The Southern Chile Trench is a southern extension of the subduction of the Antarctic and Nazca plates below South America. Heading south along the ridge, the subduction rate decreases until its remaining oblique motion evolves into the Shackleton Fracture Zone transform boundary. The south-western edge of the plate is bounded by the Shetland microplate separating the Shackleton Fracture Zone and the South Scotia Ridge.

North of South Shetland Islands and along the southern half of the Shackleton Fracture Zone is the remnant of the Phoenix Plate (also known as Drake or Aluk Plate). Around 47 Ma the subduction of the Phoenix Plate started as the propagation of the Pacific-Antarctic Ridge continued. The last collision between Phoenix ridge segments and the subduction zone was 6.5 Ma and at 3.3 Ma movements had stopped and the remnants of the Phoenix Plate was incorporated into the Antarctic Plate. The southern part of the Shackleton Fracture Zone is the former eastern edge of the Phoenix Plate.

East Scotia Ridge

The eastern edge of the Scotia plate is a spreading ridge bounded by the South Sandwich microplate, forming the East Scotia Ridge. The East Scotia Ridge is a back-arc spreading ridge that formed due to subduction of the South American plate below the South Sandwich plate along the South Sandwich Island arc. Exact spreading rates are still being disputed in the literature, but it has been agreed that rates range between 60–90 mm/yr.

The banks of northern Central Scotia Sea are superposed on oceanic basement and the spreading centre of the West Scotia Sea. Analyses of samples of volcaniclastic rocks from these sites indicate they are constructs of a continental arc and in some cases oceanic arc similar to those being formed in the currently active South Sandwich Arc. The oldest volcanic arc activity in the central and eastern regions of the Scotia Sea are 28.5 Ma. The South Sandwich forarc originated in the Central Scotia Sea at that time but has since been translated eastward by the back-arc spreading centre of the East Scotia Ridge.

Time-line

The timing of the formation of the Scotia plate and opening of the Drake Passage have long been the subject of much debate due to the important implications for changes in ocean currents and shifts in paleoclimate. The thermal isolation of Antarctica, engendering the formation of the Antarctic ice sheet, has largely been attributed to the opening of the Drake Passage.

Formation

The Scotia Plate originated about 80 million years ago (Ma), during the late Mesozoic at the Panthalassic margins of the Gondwana supercontinent between two Precambrian cratons, the Kalahari and East Antarctic Cratons, now located in Africa and Antarctica. Its development was also influenced by the Río de la Plata Craton in South America. The initial cause for its formation was the break-up of Gondwana in what became the south-west Indian Ocean.

The earliest marine fossils found on the Maurice Ewing Bank, on the eastern end of the Falkland Plateau, are associated with the Indian and Tethys Oceans and probably slightly more than 150 Ma. The Weddell Sea opened and spread along the southern margin of the Falkland Plateau and into the Rocas Verdes back-arc basin which extends from South Georgia and along the Patagonian Andes. Fragments of the inverted oceanic basement of this basin are preserved as ophiolitic complexes in this area, including the Larsen Harbour complex on South Georgia. This makes it possible to restore the original position of the South Georgia microcontinent south of the Burdwood Bank on the western North Scotia Ridge south of the Falkland Islands.

The Rocas Verdes basin was filled with turbidites derived from the volcanic arc on its Pacific margin and partly from its continental margin. The Weddell Sea continued to expand which led to the extension between the Patagonian Andes and the Antarctic Peninsula. In the Mid-Cretaceous (100 Ma) the spreading rate in South

Atlantic increased significantly and the Mid-Atlantic Ridge grew from 1200 km to 7000 km. This led to compressional deformations along the western margin of the South American Plate and the obduction of the Rocas Verdes basement onto this margin. Structures associated with this obduction are found from Tierra del Fuego to South Georgia. The acceleration of the westward motion of South America and the inversion of the Rocas Verde Basin finally lead to the initiation of the Scotia Arc. This inversion had a strike-slip component which can be seen in the Cooper Bay dislocation on South Georgia. This geological regime lead to the uplift and elongation of the Andes and the embryonic North Scotia Ridge, which resulted in the initial eastward relocation of the South Georgia microcontinent and formation of the Central Scotia Sea.

Opening of Drake Passage

Between the Late Cretaceous and the Early Oligocene (90-30 Ma) little changed in the region, except for the subduction of the Phoenix Plate along the Shackleton Fracture Zone. The Late Paleocene and Early Eocene (60-50 Ma) saw the formation of South Scotia Sea and South Scotia Ridge — the first sign of separation of the southern Andes and the Antarctic Peninsula — which resulted in seafloor spreading in the West Scotia Sea and hence the initial opening of a deep Drake Passage. The ongoing lengthening of the North Scotia Rridge beyond the Burdwood Bank caused South Georgia to move further east. The banks of the North Scotia Ridge contain volcanic rocks similar to those found in Tierra del Fuego, including on Isla de los Estados on the easternmost tip.

During the early Eocene (50 Ma), the Drake Passage between the southern tip of South America at Cape Horn and the South Shetland Islands of Antarctica was a small opening with limited circulation. A change in relative motion between the South American plate and the Antarctic plate would have severe effects, causing seafloor spreading and the formation of the Scotia plate. Marine geophysical data indicates that motion between the South American plate and the Antarctic plate shifted from N-S to WNW-ESE accompanied by an eightfold increase in the separation rate. This shift in spreading initiated crustal thinning and by 30-34 Ma, the West Scotia Ridge formed.

One of the most prominent features in the Scotia plate itself is the median valley known as the West Scotia Ridge. It was produced by two plates that are no longer independently active: the Magallanes and Central Scotia Plates. The ridge consists of seven segments separated by right-lateral transform offsets of various lengths. The western region of the Scotia Plate can be dated to 26–5.5 Ma. A complex pattern of spreading prior to 26 Ma is probably present in the oldest parts of this spreading regime; i.e. west of Terror Rise (north of Elephant Island) and on the shelf slope of Tierra del Fuego. Until 17 Ma the Central Scotia Plate moved quickly eastwards, fuelled by the eastern trench migration, but both plates have moved very slowly since.

References

- Frisch, Wolfgang; Meschede, Martin; Blakey, Ronald C. (2010), Plate Tectonics: Continental Drift and Mountain Building, Springer, pp. 11–12, ISBN 9783540765042.

- Hillis, R. R.; Müller, R. D. (2003). Evolution and Dynamics of the Australian Plate. Boulder, CO: Geological Society of America. p. 363. ISBN 0-8137-2372-8.

- Feldman, Jay (2005). When the Mississippi Ran Backwards : Empire, Intrigue, Murder, and the New Madrid Earthquakes. Free Press. ISBN 0-7432-4278-5.

- Pisco, Peru, Earthquake of August 15, 2007: Lifeline Performance. Reston, VA: ASCE, Technical Council on Lifeline Earthquake Engineering. ISBN 9780784410615.

- Wohletz, K.H.; Brown, W.K. "SFT and the Earth's Tectonic Plates". Los Alamos National Laboratory. Archived from the original on February 17, 2013.

4

Crust: An Integrated Study

The outer shell of any planet is termed as the crust. It is majorly formed by igneous processes and it occupies less than 1% of the Earth's volume. Some other aspects elucidated in the section are oceanic crust, asthenosphere, mantle and lithosphere. The chapter offers an insightful focus on the topic and incorporates its major aspects.

Crust (Geology)

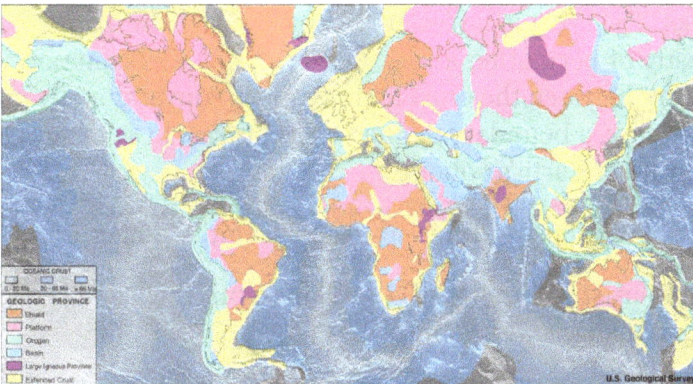

Geologic provinces of the World (USGS)

Shield

Platform

Orogen

Basin

Large igneous province

Extended crust

Oceanic crust:

0–20 Ma

20–65 Ma

>65 Ma

In geology, the crust is the outermost solid shell of a rocky planet or natural satellite, which is chemically distinct from the underlying mantle. The crusts of Earth, the Moon, Mercury, Venus, Mars, Io, and other planetary bodies have been generated largely by igneous processes, and these crusts are richer in incompatible elements than their respective mantles.

Earth's Crust

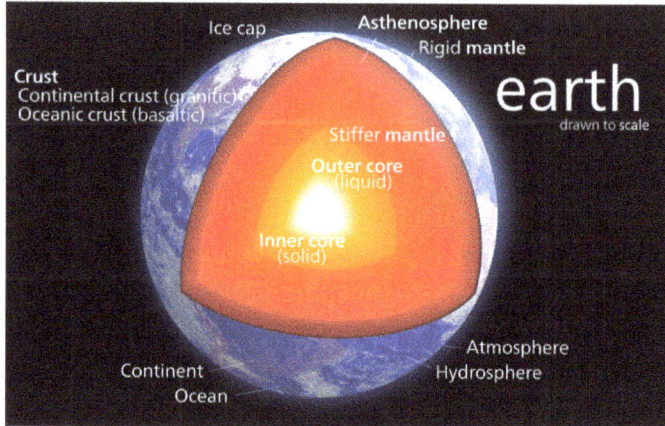

The internal structure of Earth

The crust of the Earth is composed of a great variety of igneous, metamorphic, and sedimentary rocks. The crust is underlain by the mantle. The upper part of the mantle is composed mostly of peridotite, a rock denser than rocks common in the overlying crust. The boundary between the crust and mantle is conventionally placed at the Mohorovičić discontinuity, a boundary defined by a contrast in seismic velocity. The crust occupies less than 1% of Earth's volume.

The oceanic crust of the sheet is different from its continental crust.

- The oceanic crust is 5 km (3 mi) to 10 km (6 mi) thick and is composed primarily of basalt, diabase, and gabbro.

- The continental crust is typically from 30 km (20 mi) to 50 km (30 mi) thick and is mostly composed of slightly less dense rocks than those of the oceanic crust. Some of these less dense rocks, such as granite, are common in the continental crust but rare to absent in the oceanic crust.

Both the continental and oceanic crust "float" on the mantle. Because the continental crust is thicker, it extends both to greater elevations and greater depth than the oceanic crust. The slightly lower density of felsic continental rock compared to basaltic oceanic rock contributes to the higher relative elevation of the top of the continental crust. As the top of the continental crust reaches elevations higher than that of the oceanic, water runs off the continents and collects above the oceanic crust. Because of the change in velocity of seismic waves it is believed that beneath continents at a certain depth continental crust (sial) becomes close in its physical properties to oceanic crust (sima), and the transition zone is referred to as the Conrad discontinuity.

The temperature of the crust increases with depth, reaching values typically in the range from about 200 °C (392 °F) to 400 °C (752 °F) at the boundary with the underlying mantle. The crust and underlying relatively rigid uppermost mantle make up the lithosphere. Because of convection in the underlying plastic (although non-molten) upper mantle and asthenosphere, the lithosphere is broken into tectonic plates that move. The temperature increases by as much as 30 °C (about 50 °F) for every kilometer locally in the upper part of the crust, but the geothermal gradient is smaller in deeper crust.

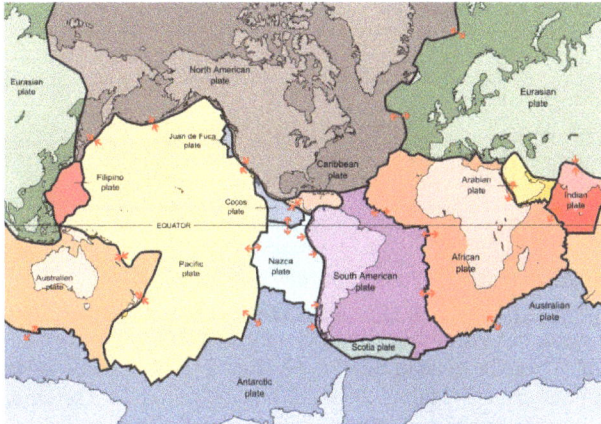

Plates in the crust of Earth

Partly by analogy to what is known about the Moon, Earth is considered to have differentiated from an aggregate of planetesimals into its core, mantle and crust within about 100 million years of the formation of the planet, 4.6 billion years ago. The primordial crust was very thin and was probably recycled by much more vigorous plate tectonics and destroyed by significant asteroid impacts, which were much more common in the early stages of the solar system.

Earth has probably always had some form of basaltic crust, but the age of the oldest oceanic crust today is only about 200 million years. In contrast, the bulk of the continental crust is much older. The oldest continental crustal rocks on Earth have ages in the range from about 3.7 to 4.28 billion years and have been found in the Narryer Gneiss Terrane in Western Australia, in the Acasta Gneiss in the Northwest Territories on the Canadian Shield, and on other cratonic regions such as those on the Fennoscandian Shield. Some zircon with age as great as 4.3 billion years has been found in the Narryer Gneiss Terrane.

The average age of the current Earth's continental crust has been estimated to be about 2.0 billion years. Most crustal rocks formed before 2.5 billion years ago are located in cratons. Such old continental crust and the underlying mantle asthenosphere are less dense than elsewhere in Earth and so are not readily destroyed by subduction. Formation of new continental crust is linked to periods of intense orogeny; these periods coincide with the formation of the supercontinents such as Rodinia, Pangaea and Gondwana. The crust forms in part by aggregation of island arcs including granite and metamorphic fold belts, and it is preserved in part by depletion of the underlying mantle to form buoyant lithospheric mantle.

Composition

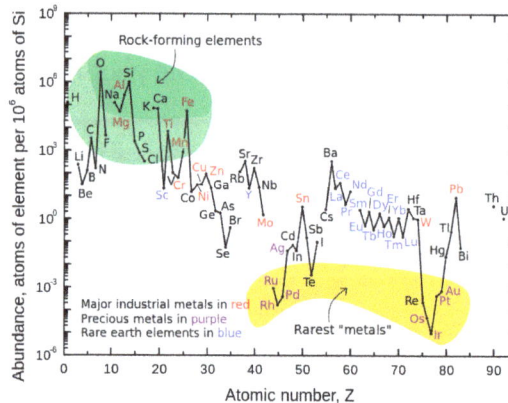

Abundance (atom fraction) of the chemical elements in Earth's
upper continental crust as a function of atomic number. The rarest elements in the crust (shown in
yellow) are not the heaviest, but are rather the siderophile (iron-loving) elements in the Goldschmidt
classification of elements. These have been depleted by being relocated deeper into Earth's core. Their
abundance in meteoroid materials is higher. Additionally, tellurium and selenium have been depleted
from the crust due to formation of volatile hydrides.

The continental crust has an average composition similar to that of andesite. Continental crust is enriched in incompatible elements compared to the basaltic ocean crust and much enriched compared to the underlying mantle. Although the continental crust comprises only about 0.6 weight percent of the silicate on Earth, it contains 20% to 70% of the incompatible elements.

Most Abundant Elements of Earth's Crust	Approximate % by weight
O	46.6
Si	27.7
Al	8.1
Fe	5.0
Ca	3.6
Na	2.8
K	2.6
Mg	1.5

Oxide	Percent
SiO_2	60.6
Al_2O_3	15.9
CaO	6.4

Oxide	Percent
MgO	4.7
Na_2O	3.1
Fe as FeO	6.7
K_2O	1.8
TiO_2	0.7
P_2O_5	0.1

All the other constituents except water occur only in very small quantities and total less than 1%. Estimates of average density for the upper crust range between 2.69 and 2.74 g/cm^3 and for lower crust between 3.0 and 3.25 g/cm^3.

Moon's Crust

A theoretical protoplanet named "Theia" is thought to have collided with the forming Earth, and part of the material ejected into space by the collision accreted to form the Moon. As the Moon formed, the outer part of it is thought to have been molten, a "lunar magma ocean." Plagioclase feldspar crystallized in large amounts from this magma ocean and floated toward the surface. The cumulate rocks form much of the crust. The upper part of the crust probably averages about 88% plagioclase (near the lower limit of 90% defined for anorthosite): the lower part of the crust may contain a higher percentage of ferromagnesian minerals such as the pyroxenes and olivine, but even that lower part probably averages about 78% plagioclase. The underlying mantle is denser and olivine-rich.

The thickness of the crust ranges between about 20 and 120 km. Crust on the far side of the Moon averages about 12 km thicker than that on the near side. Estimates of average thickness fall in the range from about 50 to 60 km. Most of this plagioclase-rich crust formed shortly after formation of the moon, between about 4.5 and 4.3 billion years ago. Perhaps 10% or less of the crust consists of igneous rock added after the formation of the initial plagioclase-rich material. The best-characterized and most voluminous of these later additions are the mare basalts formed between about 3.9 and 3.2 billion years ago. Minor volcanism continued after 3.2 billion years, perhaps as recently as 1 billion years ago. There is no evidence of plate tectonics.

Study of the Moon has established that a crust can form on a rocky planetary body significantly smaller than Earth. Although the radius of the Moon is only about a quarter that of Earth, the lunar crust has a significantly greater average thickness. This thick crust formed almost immediately after formation of the Moon. Magmatism continued after the period of intense meteorite impacts ended about 3.9 billion years ago, but igneous rocks younger than 3.9 billion years make up only a minor part of the crust.

Structure of the Earth

The interior structure of the Earth is layered in spherical shells, like an onion. These layers can be defined by their chemical and their rheological properties. Earth has an outer silicate solid crust, a highly viscous mantle, a liquid outer core that is much less viscous than the mantle, and a solid inner core. Scientific understanding of the internal structure of the Earth is based on observations of topography and bathymetry, observations of rock in outcrop, samples brought to the surface from greater depths by volcanoes or volcanic activity, analysis of the seismic waves that pass through the Earth, measurements of the gravitational and magnetic fields of the Earth, and experiments with crystalline solids at pressures and temperatures characteristic of the Earth's deep interior.

Mass

The force exerted by Earth's gravity can be used to calculate its mass. Astronomers can also calculate Earth's mass by observing the motion of orbiting satellites. Earth's average density can be determined through gravitometric experiments, which have historically involved pendulums.

The mass of Earth is about 6×10^{24} kg.

Structure

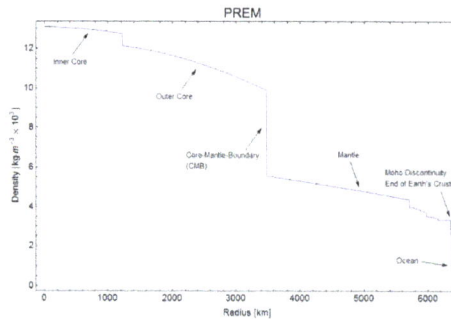

Earth's radial density distribution according to the preliminary reference earth model (PREM).

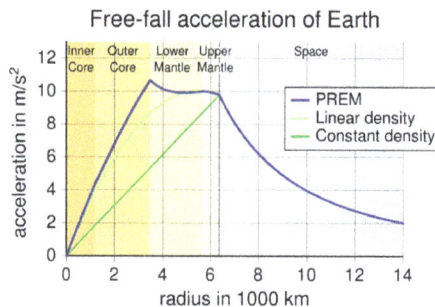

Earth's gravity according to the preliminary reference earth model (PREM).
Comparison to approximations using constant and linear density for Earth's interior.

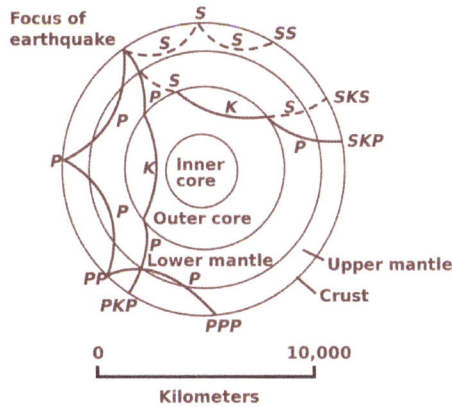

Mapping the interior of Earth with earthquake waves.

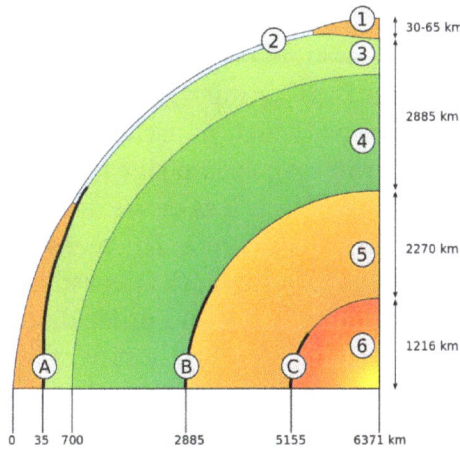

Schematic view of the interior of Earth. 1. continental crust – 2. oceanic crust – 3. upper mantle – 4. lower mantle – 5. outer core – 6. inner core – A: Mohorovičić discontinuity – B: Gutenberg Discontinuity – C: Lehmann–Bullen discontinuity.

The structure of Earth can be defined in two ways: by mechanical properties such as rheology, or chemically. Mechanically, it can be divided into lithosphere, asthenosphere, mesospheric mantle, outer core, and the inner core. Chemically, Earth can be divided into the crust, upper mantle, lower mantle, outer core, and inner core. The geologic component layers of Earth are at the following depths below the surface:

Depth		Layer
Kilometres	Miles	
0–60	0–37	Lithosphere (locally varies between 5 and 200 km)
0–35	0–22	... Crust (locally varies between 5 and 70 km)
35–60	22–37	... Uppermost part of mantle
35–2,890	22–1,790	Mantle
210-270	130-168	... Upper mesosphere (upper mantle)

660–2,890	410–1,790	... Lower mesosphere (lower mantle)
2,890–5,150	1,790–3,160	Outer core
5,150–6,360	3,160–3,954	Inner core

The layering of Earth has been inferred indirectly using the time of travel of refracted and reflected seismic waves created by earthquakes. The core does not allow shear waves to pass through it, while the speed of travel (seismic velocity) is different in other layers. The changes in seismic velocity between different layers causes refraction owing to Snell's law, like light bending as it passes through a prism. Likewise, reflections are caused by a large increase in seismic velocity and are similar to light reflecting from a mirror.

Crust

The crust ranges from 5–70 km (~3–44 miles) in depth and is the outermost layer. The thin parts are the oceanic crust, which underlie the ocean basins (5–10 km) and are composed of dense (mafic) iron magnesium silicate igneous rocks, like basalt. The thicker crust is continental crust, which is less dense and composed of (felsic) sodium potassium aluminium silicate rocks, like granite. The rocks of the crust fall into two major categories – sial and sima (Suess,1831–1914). It is estimated that sima starts about 11 km below the Conrad discontinuity (a second order discontinuity). The uppermost mantle together with the crust constitutes the lithosphere. The crust-mantle boundary occurs as two physically different events. First, there is a discontinuity in the seismic velocity, which is most commonly known as the Mohorovičić discontinuity or Moho. The cause of the Moho is thought to be a change in rock composition from rocks containing plagioclase feldspar (above) to rocks that contain no feldspars (below). Second, in oceanic crust, there is a chemical discontinuity between ultramafic cumulates and tectonized harzburgites, which has been observed from deep parts of the oceanic crust that have been obducted onto the continental crust and preserved as ophiolite sequences.

Many rocks now making up Earth's crust formed less than 100 million (1×10^8) years ago; however, the oldest known mineral grains are 4.4 billion (4.4×10^9) years old, indicating that Earth has had a solid crust for at least that long.

Mantle

Earth's mantle extends to a depth of 2,890 km, making it the thickest layer of Earth. The mantle is divided into upper and lower mantle. The upper and lower mantle are separated by the transition zone. The lowest part of the mantle next to the core-mantle boundary is known as the D″ (pronounced dee-double-prime) layer. The pressure at the bottom of the mantle is ~140 GPa (1.4 Matm). The mantle is composed of silicate rocks that are rich in iron and magnesium relative to the overlying crust. Although solid, the high temperatures within the mantle cause the silicate material to be sufficiently ductile that it can flow on very long timescales. Convection of the mantle is expressed at the surface through the motions of tectonic plates. As there is intense and increasing pressure as one travels deeper into the mantle, the lower part of the mantle flows less easily than does the upper mantle (chemical changes within the mantle may also be important). The viscosity of the mantle ranges between 10^{21} and 10^{24} Pa·s, depending on depth. In comparison, the viscosity of water is approximately 10^{-3} Pa·s and that of pitch is 10^{7} Pa·s. The source of heat that drives plate tectonics is the primordial heat left over from the planet's formation as well as the radioactive decay of uranium, thorium, and potassium in Earth's crust and mantle.

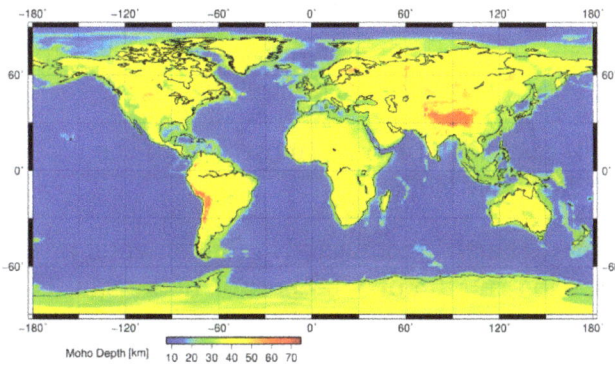

World map showing the position of the Moho.

Core

The average density of Earth is 5,515 kg/m³. Because the average density of surface material is only around 3,000 kg/m³, we must conclude that denser materials exist within Earth's core. Seismic measurements show that the core is divided into two parts, a "solid" inner core with a radius of ~1,220 km and a liquid outer core extending beyond it to a radius of ~3,400 km. The densities are between 9,900 and 12,200 kg/m³ in the outer core and 12,600–13,000 kg/m³ in the inner core.

The inner core was discovered in 1936 by Inge Lehmann and is generally believed to be composed primarily of iron and some nickel. It is not necessarily a solid, but, because it is able to deflect seismic waves, it must behave as a solid in some fashion. Experimental evidence has at times been critical of crystal models of the core. Other experimental studies show a discrepancy under high pressure: diamond anvil (static) studies at core pressures yield melting temperatures that are approximately 2000 K below those from shock laser (dynamic) studies. The laser studies create plasma, and the results are suggestive that constraining inner core conditions will depend on whether the inner core is a solid or is a plasma with the density of a solid. This is an area of active research.

In early stages of Earth's formation about four and a half billion (4.5×10^{9}) years ago, melting would have caused denser substances to sink toward the center in a process called planetary differentiation, while less-dense materials would have migrated to the crust. The core is thus believed to largely be composed of iron (80%), along with nickel and one or more light elements, whereas other dense elements, such as lead and uranium, either are too rare to be significant or tend to bind to lighter elements and thus remain in the crust. Some have argued that the inner core may be in the form of a single iron crystal.

Under laboratory conditions a sample of iron–nickel alloy was subjected to the corelike pressures by gripping it in a vise between 2 diamond tips (diamond anvil cell), and then heating to approximately 4000 K. The sample was observed with x-rays, and strongly supported the theory that Earth's inner core was made of giant crystals running north to south.

The liquid outer core surrounds the inner core and is believed to be composed of iron mixed with nickel and trace amounts of lighter elements.

Recent speculation suggests that the innermost part of the core is enriched in gold, platinum and other siderophile elements.

The matter that comprises Earth is connected in fundamental ways to matter of certain chondrite meteorites, and to matter of outer portion of the Sun. There is good reason to believe that Earth is, in the main, like a chondrite meteorite. Beginning as early as 1940, scientists, including Francis Birch, built geophysics upon the premise that Earth is like ordinary chondrites, the most common type of meteorite observed impacting Earth, while totally ignoring another, albeit less abundant type, called enstatite chondrites. The principal difference between the two meteorite types is that enstatite chondrites formed under circumstances of extremely limited available oxygen, leading to certain normally oxyphile elements existing either partially or wholly in the alloy portion that corresponds to the core of Earth.

Dynamo theory suggests that convection in the outer core, combined with the Coriolis effect, gives rise to Earth's magnetic field. The solid inner core is too hot to hold a permanent magnetic field but probably acts to stabilize the magnetic field generated by the liquid outer core. The average magnetic field strength in Earth's outer core is estimated to be 25 Gauss (2.5 mT), 50 times stronger than the magnetic field at the surface.

Recent evidence has suggested that the inner core of Earth may rotate slightly faster than the rest of the planet; however, more recent studies in 2011 found this hypothesis to be inconclusive. Options remain for the core which may be oscillatory in nature or a chaotic system. In August 2005 a team of geophysicists announced in the journal *Science* that, according to their estimates, Earth's inner core rotates approximately 0.3 to 0.5 degrees per year faster relative to the rotation of the surface.

The current scientific explanation for Earth's temperature gradient is a combination of heat left over from the planet's initial formation, decay of radioactive elements, and freezing of the inner core.

Historical Development of Alternative Conceptions

Edmond Halley's hypothesis.

In 1692 Edmond Halley (in a paper printed in *Philosophical Transactions of Royal Society of London*) put forth the idea of Earth consisting of a hollow shell about 500 miles thick, with two inner concentric shells around an innermost core, corresponding to the diameters of the planets Venus, Mars, and Mercury respectively. Halley's construct was a method of accounting for the (flawed) values of the relative density of Earth and the Moon that had been given by Sir Isaac Newton, in *Principia* (1687). "Sir Isaac Newton has demonstrated the Moon to be more solid than our Earth, as 9 to 5," Halley remarked; "why may we not then suppose four ninths of our globe to be cavity?"

Oceanic Crust

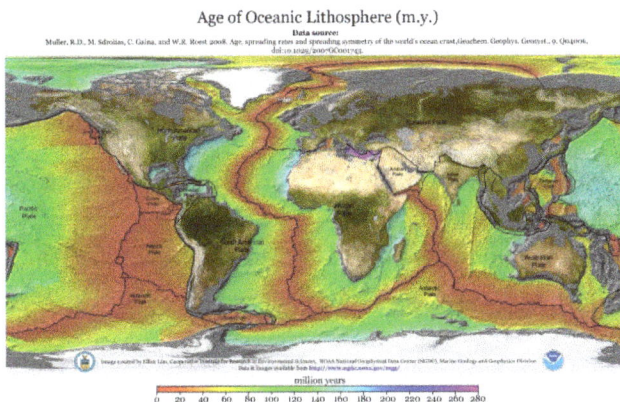

Colors indicate the age of oceanic lithosphere, wherein red indicates the youngest age, and blue indicates the oldest age. The lines represent tectonic plates.

Oceanic crust is the uppermost layer of the oceanic portion of a tectonic plate. The crust overlies the solidified and uppermost layer of the mantle. The crust and the solid mantle layer together constitute oceanic lithosphere.

Oceanic crust is the result of erupted mantle material originating from below the plate, cooled and in most instances, modified chemically by seawater. This occurs mostly at mid-ocean ridges, but also at scattered hotspots, and also in rare but powerful occurrences known as flood basalt eruptions. It is primarily composed of mafic rocks, or sima, which is rich in iron and magnesium. It is thinner than continental crust, or sial, generally less than 10 kilometers thick; however it is denser, having a mean density of about 2.9 grams per cubic centimeter as opposed to continental crust which has a density of about 2.7 grams per cubic centimeter.

Composition

Although a complete section of oceanic crust has not yet been drilled, geologists have several pieces of evidence that help them understand the ocean floor. Estimations of composition are based on analyses of ophiolites (sections of oceanic crust that are preserved on the continents), comparisons of the seismic structure of the oceanic crust with laboratory determinations of seismic velocities in known rock types, and samples recovered from the ocean floor by submersibles, dredging (especially from ridge crests and fracture zones) and drilling. Oceanic crust is significantly simpler than continental crust and generally can be divided in three layers.

- Layer 1 is on an average 0.4 km thick. It consists of unconsolidated or semiconsolidated sediments, usually thin or even not present near the mid-ocean ridges but thickens farther away from the ridge. Near the continental margins sediment is terrigenous, meaning derived from the land, unlike deep sea sediments which are made of tiny shells of marine organisms, usually calcareous and siliceous, or it can be made of volcanic ash and terrigenous sediments transported by turbidity currents.

- Layer 2 could be divided into two parts: layer 2A – 0.5 km thick uppermost volcanic layer of glassy to finely crystalline basalt usually in the form of pillow basalt, and layer 2B – 1.5 km thick layer composed of diabase dikes.

- Layer 3 is formed by slow cooling of magma beneath the surface and consists of coarse grained gabbros and cumulate ultramafic rocks. It constitutes over two-thirds of oceanic crust volume with almost 5 km thickness.

Geochemistry

The most voluminous volcanic rocks of the ocean floor are the mid-oceanic ridge basalts, which are derived from low-potassium tholeiitic magmas. These rocks have low concentrations of large ion lithophile elements (LILE), light rare earth elements (LREE), volatile elements and other highly incompatible elements. There can be found basalts enriched with incompatible elements, but they are rare and associated with mid-ocean ridge hot spots such as surroundings of Galapagos Islands, the Azores and Iceland.

Life Cycle

Oceanic crust is continuously being created at mid-ocean ridges. As plates diverge at these ridges, magma rises into the upper mantle and crust. As it moves away from the ridge, the lithosphere becomes cooler and denser, and sediment gradually builds on top of it. The youngest oceanic lithosphere is at the oceanic ridges, and it gets progressively older away from the ridges.

As the mantle rises it cools and melts, as the pressure decreases and it crosses the solidus. The amount of melt produced depends only on the temperature of the mantle as it rises. Hence most oceanic crust is the same thickness (7 ± 1 km). Very slow spreading ridges (<1 cm·yr^{-1} half-rate) produce thinner crust (4–5 km thick) as the mantle has a chance to cool on upwelling and so it crosses the solidus and melts at lesser depth, thereby producing less melt and thinner crust. An example of this is the Gakkel Ridge under the Arctic Ocean. Thicker than average crust is found above plumes as the mantle is hotter and hence it crosses the solidus and melts at a greater depth, creating more melt and a thicker crust. An example of this is Iceland which has crust of thickness ~20 km.

The oceanic lithosphere subducts at what are known as convergent boundaries. These boundaries can exist between oceanic lithosphere on one plate and oceanic lithosphere on another, or between oceanic lithosphere on one plate and continental lithosphere on another. In the second situation, the oceanic lithosphere always subducts because the continental lithosphere is less dense. The subduction process consumes older oceanic lithosphere, so oceanic crust is seldom more than 200 million years old. The process of super-continent formation and destruction via repeated cycles of creation and destruction of oceanic crust is known as the Wilson cycle.

The oldest large scale oceanic crust is in the west Pacific and north-west Atlantic - both are about up to 180-200 million years old. However, parts of the eastern Mediterranean Sea are remnants of the much older Tethys ocean, at about 270 and up to 340 million years old.

Magnetic Anomalies

The *oceanic crust* displays an interesting pattern of parallel magnetic lines, parallel to the ocean ridges, frozen in the basalt. In the 1950s, scientists mapped the magnetic field generated by rocks on the ocean floor. They noticed a symmetrical pattern of positive and negative magnetic lines as they moved along the ocean floor, and the line of symmetry was at the mid ocean ridge. That the anomalies were symmetrical at the mid-ocean ridge was explained by the hypothesis that new rock was being formed by magma at the mid-ocean ridges, and the ocean floor was spreading out from this point. When the magma cooled to form rock, it aligned itself with the current position of the north magnetic pole of the Earth (which has reversed many times in its past) at the time of its cooling. New magma forced the older cooled magma away from the ridge. Approximately half of the new rock was formed on one side of the ridge and half on the other.

Continental Crust

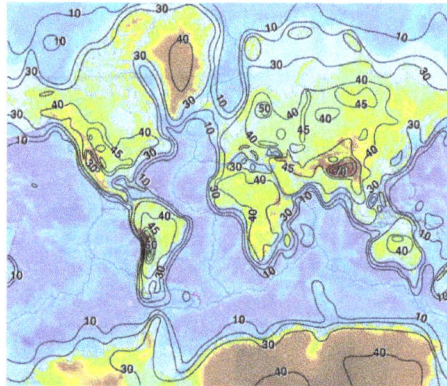

The thickness of Earth's crust (km)

The continental crust is the layer of igneous, sedimentary, and metamorphic rocks that forms the continents and the areas of shallow seabed close to their shores, known as continental shelves. This layer is sometimes called *sial* because its bulk composition is more felsic compared to the oceanic crust, called *sima* which has a more mafic bulk composition. Changes in seismic wave velocities have shown that at a certain depth (the Conrad discontinuity), there is a reasonably sharp contrast between the more felsic upper continental crust and the lower continental crust, which is more mafic in character.

The continental crust consists of various layers, with a bulk composition that is intermediate to felsic. The average density of continental crust is about 2.7 g/cm³, less dense than the ultramafic material that makes up the mantle, which has a density of around 3.3 g/cm³. Continental crust is also less dense than oceanic crust, whose density is about 2.9 g/cm³. At 25 to 70 km, continental crust is considerably thicker than oceanic crust, which has an average thickness of around 7–10 km. About 40% of Earth's surface is currently occupied by continental crust. It makes up about 70% of the volume of Earth's crust.

Importance

Because the surface of continental crust mainly lies above sea level, its existence allowed land life to evolve from marine life. Its existence also provides broad expanses of shallow water known as epeiric seas and continental shelves where complex metazoan life could become established during early Paleozoic time, in what is now called the Cambrian explosion.

Origin

There is little evidence of continental crust prior to 3.5 Ga, and there was relatively rapid development on shield areas consisting of continental crust between 3.0 and 2.5 Ga.

All continental crust ultimately derives from the fractional differentiation of oceanic crust over many eons. This process has been and continues today primarily as a result of the volcanism associated with subduction.

Forces at Work

In contrast to the persistence of continental crust, the size, shape, and number of continents are constantly changing through geologic time. Different tracts rift apart, collide and recoalesce as part of a grand supercontinent cycle. There are currently about 7 billion cubic kilometers of continental crust, but this quantity varies because of the nature of the forces involved. The relative permanence of continental crust contrasts with the short life of oceanic crust. Because continental crust is less dense than oceanic crust, when active margins of the two meet in subduction zones, the oceanic crust is typically subducted back into the mantle. Continental crust is rarely subducted (this may occur where continental crustal blocks collide and overthicken, causing deep melting under mountain belts such as the Himalayas or the Alps). For this reason the oldest rocks on Earth are within the cratons or cores of the continents, rather than in repeatedly recycled oceanic crust; the oldest intact crustal fragment is the Acasta Gneiss at 4.01 Ga, whereas the oldest oceanic crust (located on the Pacific Plate offshore of Kamchatka) is from the Jurassic (~180 Ma). Continental crust and the rock layers that lie on and within it are thus the best archive of Earth's history.

The height of mountain ranges is usually related to the thickness of crust. This results from the isostasy associated with orogeny (mountain formation). The crust is thickened by the compressive forces related to subduction or continental collision. The buoyancy of the crust forces it upwards, the forces of the collisional stress balanced by gravity and erosion. This forms a keel or mountain root beneath the mountain range, which is where the thickest crust is found. The thinnest continental crust is found in rift zones, where the crust is thinned by detachment faulting and eventually severed, replaced by oceanic crust. The edges of continental fragments formed this way (both sides of the Atlantic Ocean, for example) are termed passive margins.

The high temperatures and pressures at depth, often combined with a long history of complex distortion, cause much of the lower continental crust to be metamorphic - the main exception to this being recent igneous intrusions. Igneous rock may also be "underplated" to the underside of the crust, i.e. adding to the crust by forming a layer immediately beneath it.

Continental crust is produced and (far less often) destroyed mostly by plate tectonic processes, especially at convergent plate boundaries. Additionally, continental crustal material is transferred to oceanic crust by sedimentation. New material can be added to the continents by the partial melting of oceanic crust at subduction zones, causing the lighter material to rise as magma, forming volcanoes. Also, material can be accreted horizontally when volcanic island arcs, seamounts or similar structures collide with the side of the continent as a result of plate tectonic movements. Continental crust is also lost through erosion and sediment subduction, tectonic erosion of forearcs, delamination, and deep subduction of continental crust in collision zones. Many theories of crustal growth are controversial, including rates of crustal growth and recycling, whether the lower crust is recycled differently from the upper crust, and over how much of Earth history plate tectonics has operated and so could be the dominant mode of continental crust formation and destruction.

It is a matter of debate whether the amount of continental crust has been increasing, decreasing, or remaining constant over geological time. One model indicates that at prior to 3.7 Ga ago continental crust constituted less than 10% of the present amount. By 3.0 Ga ago the amount was about 25%, and following a period of rapid crustal evolution it was about 60% of the current amount by 2.6 Ga ago. The growth of continental crust appears to have occurred in *spurts* of increased activity corresponding to five episodes of increased production through geologic time.

Asthenosphere

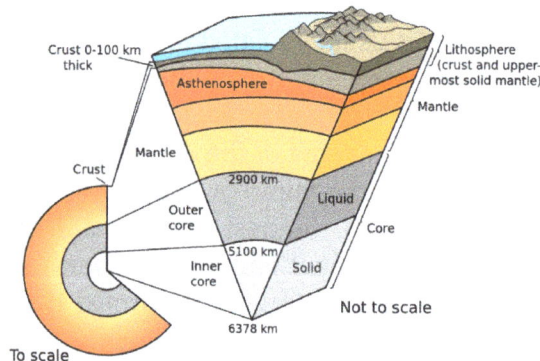

Earth cutaway from core to crust, the asthenosphere lying between the upper mantle and the lithospheric mantle (detail not to scale)

The asthenosphere is the highly viscous, mechanically weak and ductilely deforming region of the upper mantle of the Earth. It lies below the lithosphere, at depths between approximately 80 and 200 km (50 and 120 miles) below the surface. The Lithosphere-Asthenosphere boundary is usually referred to as LAB. The asthenosphere is generally solid, although some of its regions could be melted (e.g., below mid-ocean ridges). The lower boundary of the asthenosphere is not well defined. The thickness of the asthenosphere depends mainly on the temperature. In some regions the asthenosphere could extend as deep as 700 km (430 mi). It is considered the source region of mid-ocean ridge basalt (MORB).

Characteristics

The asthenosphere is a part of the upper mantle just below the lithosphere that is involved in plate tectonic movement and isostatic adjustments. The lithosphere-asthenosphere boundary is conventionally taken at the 1300 °C isotherm, above which the mantle behaves in a rigid fashion and below which it behaves in a ductile fashion. Seismic waves pass relatively slowly through the asthenosphere compared to the overlying lithospheric mantle, thus it has been called the *low-velocity zone* (LVZ), although the two are not exactly the same. This decreasing in seismic waves velocity from lithosphere to asthenosphere could be caused by the presence of small percentage of melt in the asthenosphere. The lower boundary of the LVZ lies at a depth of 180–220 km, whereas the base of the asthenosphere lies at a depth of about 700 km. This was the observation that originally alerted seismologists to its presence and gave some information about its physical properties, as the speed of seismic waves decreases with decreasing rigidity.

In the old oceanic mantle the transition from the lithosphere to the asthenosphere, the so-called lithosphere-asthenosphere boundary (LAB) is shallow (about 60 km in some regions) with a sharp and large velocity drop (5-10%). At the mid-ocean ridges the LAB rises to within a few kilometers of the ocean floor.

The upper part of the asthenosphere is believed to be the zone upon which the great rigid and brittle lithospheric plates of the Earth's crust move about. Due to the temperature and pressure conditions in the asthenosphere, rock becomes ductile, moving at rates of deformation measured in cm/yr over lineal distances eventually measuring thousands of kilometers. In this way, it flows like a convection current, radiating heat outward from the Earth's interior. Above the asthenosphere, at the same rate of deformation, rock behaves elastically and, being brittle, can break, causing faults. The rigid lithosphere is thought to "float" or move about on the slowly flowing asthenosphere, creating the movement of tectonic plates.

Historical

Although its presence was suspected as early as 1926, the worldwide occurrence of the asthenosphere was confirmed by analyses of earthquake waves from the 9.5 M_w Great Chilean earthquake of May 22, 1960.

Mantle (Geology)

The mantle is a layer inside a terrestrial planet and some other rocky planetary bodies. For a mantle to form, the planetary body must be large enough to have undergone the process of planetary differentiation by density. The mantle lies between the core below and the crust above. The terrestrial planets (Earth, Venus, Mars and Mercury), the Moon, two of Jupiter's moons (Io and Europa) and the asteroid Vesta each have a mantle made of silicate rock. Interpretation of spacecraft data suggests that at least two other moons of Jupiter (Ganymede and Callisto), as well as Titan and Triton each have a mantle made of ice or other solid volatile substances.

Earth's Mantle

The interior of Earth, similar to the other terrestrial planets, is chemically divided into layers. The mantle is a layer between the crust and the outer core. Earth's mantle is a silicate rocky shell with an average thickness of 2,886 kilometres (1,793 mi). The mantle makes up about 84% of Earth's volume. It is predominantly solid but in geological time it behaves as a very viscous fluid. The mantle encloses the hot core rich in iron and nickel, which makes up about 15% of Earth's volume. Past episodes of melting and volcanism at the shallower levels of the mantle have produced a thin crust of crystallized melt products near the surface. Information about the structure and composition of the mantle has been obtained from geophysical investigation and from direct geoscientific analyses of Earth mantle-derived xenoliths and mantle that has been exposed by mid-oceanic ridge spreading.

Two main zones are distinguished in the upper mantle: the inner asthenosphere composed of plastic flowing rock of varying thickness, on average about 200 km (120 mi) thick, and the lowermost part of the lithosphere composed of rigid rock about 50 to 120 km (31 to 75 mi) thick. A thin crust, the upper part of the lithosphere, surrounds the mantle and is about 5 to 75 km (3.1 to 46.6 mi) thick. Recent analysis of hydrous ringwoodite from the mantle suggests that there is between one and three times as much water in the transition zone between the lower and upper mantle than in all the world's oceans combined.

In some places under the ocean the mantle is actually exposed on the surface of Earth. There are also a few places on land where mantle rock has been pushed to the surface by tectonic activity, most notably the Tablelands region of Gros Morne National Park in the Canadian province of Newfoundland and Labrador and St. John's Island, Egypt or Zabargad in the Red Sea. (Also Troodos Ophiolite, Lizard complex, Semail Ophiolite, and other Ophiolites)

Structure

The mantle is divided into sections which are based upon results from seismology. These layers (and their thicknesses/depths) are the following: the upper mantle (starting at the Moho, or base of the crust around 7 to 35 km (4.3 to 21.7 mi) downward to 410 km (250 mi)), the transition zone (410–660 km or 250–410 mi), the lower mantle (660–2,891 km or 410–1,796 mi), and anomalous core–mantle boundary with a variable thickness (on average ~200 km (120 mi) thick).

The top of the mantle is defined by a sudden increase in seismic velocity, which was first noted by Andrija Mohorovičić in 1909; this boundary is now referred to as the Mohor-ovičić discontinuity or "Moho". The uppermost mantle plus overlying crust are relative-ly rigid and form the lithosphere, an irregular layer with a maximum thickness of per-haps 200 km (120 mi). Below the lithosphere the upper mantle becomes notably more plastic. In some regions below the lithosphere, the seismic shear velocity is reduced; this so-called low-velocity zone (LVZ) extends down to a depth of several hundred km. Inge Lehmann discovered a seismic discontinuity at about 220 km (140 mi) depth; although this discontinuity has been found in other studies, it is not known whether the discontinuity is ubiquitous. The transition zone is an area of great complexity; it physically separates the upper and lower mantle. Very little is known about the lower mantle apart from that it appears to be relatively seismically homogeneous. The D" lay-er at the core–mantle boundary separates the mantle from the core. In 2015, research using gravitational data from GRACE satellites and the long wavelength nonhydrostat-ic geoid indicated viscosity increases by a factor of ten to 150 about 1,000 kilometres (620 mi) below earth's surface; separate research also indicates sinking tectonic plates stall at this depth, leading Robert van der Hilst to speculate "In term's of structure and dynamics, 1,000 kilometers could be more important" (than the currently accepted 660 km depth upper—lower division). The lower mantle also contains some discon-tinuous zones, called "thermochemical piles" which have been interpreted as either thermally differentiated, upwellings bringing warmer material towards the surface, or as chemically differentiated material. A principal source of the heat that drives plate tectonics is the radioactive decay of uranium, thorium, and potassium in Earth's crust and mantle.

Characteristics

The mantle differs substantially from the crust in its mechanical properties which is the direct consequence of chemical composition change (expressed as different min-eralogy). The distinction between crust and mantle is based on chemistry, rock types, rheology and seismic characteristics. The crust is a solidification product of mantle derived melts, expressed as various degrees of partial melting products during geologic time. Partial melting of mantle material is believed to cause incompatible elements to separate from the mantle, with less dense material floating upward through pore spac-es, cracks, or fissures, that would subsequently cool and solidify at the surface. Typical mantle rocks have a higher magnesium to iron ratio and a smaller proportion of silicon and aluminium than the crust. This behavior is also predicted by experiments that part-ly melt rocks thought to be representative of Earth's mantle.

Mantle rocks shallower than about 410 km (250 mi) depth consist mostly of olivine, py-roxenes, spinel-structure minerals, and garnet; typical rock types are thought to be per-idotite, dunite (olivine-rich peridotite), and eclogite. Between about 400 km (250 mi) and 650 km (400 mi) depth, olivine is not stable and is replaced by high pressure poly-morphs with approximately the same composition: one polymorph is wadsleyite (also called *beta-spinel* type), and the other is ringwoodite (a mineral with the *gamma-spinel* structure). Below about 650 km (400 mi), all of the minerals of the upper mantle begin to become unstable. The most abundant minerals present, the silicate perovskites, have structures (but not compositions) like that of the mineral perovskite followed by the magnesium/iron oxide ferropericlase. The changes in mineralogy at about 400 and 650 km (250 and 400 mi) yield distinctive signatures in seismic records of the Earth's interior, and like the moho, are readily detected using seismic waves. These changes in mineralogy may influence mantle convection, as they result in density changes and they may absorb or release latent heat as well as depress or elevate the depth of the polymorphic phase transitions for regions of different temperatures. The changes in mineralogy with depth have been investigated by laboratory experiments that duplicate high mantle pressures, such as those using the diamond anvil.

Composition of Earth's mantle in weight percent

Element	Amount	Compound	Amount
O	44.8		
Mg	22.8	SiO_2	46
Si	21.5	MgO	37.8
Fe	5.8	FeO	7.5
Ca	2.3	Al_2O_3	4.2
Al	2.2	CaO	3.2
Na	0.3	Na_2O	0.4
K	0.03	K_2O	0.04
Sum	99.7	Sum	99.1

The inner core is solid, the outer core is liquid, and the mantle solid/plastic. This is because of the relative melting points of the different layers (nickel–iron core, silicate crust and mantle) and the increase in temperature and pressure as depth increases. At the surface both nickel–iron alloys and silicates are sufficiently cool to be solid. In the upper mantle, the silicates are generally solid (localised regions with small amounts of melt exist); however, as the upper mantle is both hot and under relatively little pressure, the rock in the upper mantle has a relatively low viscosity. In contrast, the lower mantle is under tremendous pressure and therefore has a higher viscosity than the upper mantle. The metallic nickel–iron outer core is liquid because of the high temperature, despite the high pressure. As the pressure increases, the nickel–iron inner core becomes solid because the melting point of iron increases dramatically at these high pressures.

Temperature

In the mantle, temperatures range between 500 to 900 °C (932 to 1,652 °F) at the upper boundary with the crust; to over 4,000 °C (7,230 °F) at the boundary with the core. Although the higher temperatures far exceed the melting points of the mantle rocks at the surface (about 1200 °C for representative peridotite), the mantle is almost exclusively solid. The enormous lithostatic pressure exerted on the mantle prevents melting, because the temperature at which melting begins (the solidus) increases with pressure.

Movement

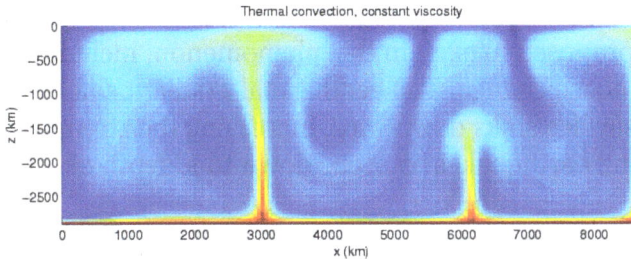

This figure is a snapshot of one time-step in a model of mantle convection. Colors closer to red are hot areas and colors closer to blue are cold areas. In this figure, heat received at the core–mantle boundary results in thermal expansion of the material at the bottom of the model, reducing its density and causing it to send plumes of hot material upwards. Likewise, cooling of material at the surface results in its sinking.

Because of the temperature difference between the Earth's surface and outer core and the ability of the crystalline rocks at high pressure and temperature to undergo slow, creeping, viscous-like deformation over millions of years, there is a convective material circulation in the mantle. Hot material upwells, while cooler (and heavier) material sinks downward. Downward motion of material occurs at convergent plate boundaries called subduction zones. Locations on the surface that lie over plumes are predicted to have high elevation (because of the buoyancy of the hotter, less-dense plume beneath) and to exhibit hot spot volcanism. The volcanism often attributed to deep mantle plumes is alternatively explained by passive extension of the crust, permitting magma to leak to the surface (the "Plate" hypothesis).

The convection of the Earth's mantle is a chaotic process (in the sense of fluid dynamics), which is thought to be an integral part of the motion of plates. Plate motion should not be confused with continental drift which applies purely to the movement of the crustal components of the continents. The movements of the lithosphere and the underlying mantle are coupled since descending lithosphere is an essential component of convection in the mantle. The observed continental drift is a complicated relationship between the forces causing oceanic lithosphere to sink and the movements within Earth's mantle.

Although there is a tendency to larger viscosity at greater depth, this relation is far from linear and shows layers with dramatically decreased viscosity, in particular in the upper mantle and at the boundary with the core. The mantle within about 200 km (120 mi) above the core–mantle boundary appears to have distinctly different seismic properties than the mantle at slightly shallower depths; this unusual mantle region just above the core is called D″ ("D double-prime"), a nomenclature introduced over 50 years ago by the geophysicist Keith Bullen. D″ may consist of material from subducted slabs that descended and came to rest at the core–mantle boundary and/or from a new mineral polymorph discovered in perovskite called post-perovskite.

Earthquakes at shallow depths are a result of stick-slip faulting; however, below about 50 km (31 mi) the hot, high pressure conditions ought to inhibit further seismicity. The mantle is considered to be viscous and incapable of brittle faulting. However, in subduction zones, earthquakes are observed down to 670 km (420 mi). A number of mechanisms have been proposed to explain this phenomenon, including dehydration, thermal runaway, and phase change. The geothermal gradient can be lowered where cool material from the surface sinks downward, increasing the strength of the surrounding mantle, and allowing earthquakes to occur down to a depth of 400 km (250 mi) and 670 km (420 mi).

The pressure at the bottom of the mantle is ~136 GPa (1.4 million atm). Pressure increases as depth increases, since the material beneath has to support the weight of all the material above it. The entire mantle, however, is thought to deform like a fluid on long timescales, with permanent plastic deformation accommodated by the movement of point, line, and/or planar defects through the solid crystals comprising the mantle. Estimates for the viscosity of the upper mantle range between 10^{19} and 10^{24} Pa·s, depending on depth, temperature, composition, state of stress, and numerous other factors. Thus, the upper mantle can only flow very slowly. However, when large forces are applied to the uppermost mantle it can become weaker, and this effect is thought to be important in allowing the formation of tectonic plate boundaries.

Exploration

Exploration of the mantle is generally conducted at the seabed rather than on land because of the relative thinness of the oceanic crust as compared to the significantly thicker continental crust.

The first attempt at mantle exploration, known as Project Mohole, was abandoned in 1966 after repeated failures and cost over-runs. The deepest penetration was approximately 180 m (590 ft). In 2005 an oceanic borehole reached 1,416 metres (4,646 ft) below the sea floor from the ocean drilling vessel *JOIDES Resolution*.

On 5 March 2007, a team of scientists on board the RRS *James Cook* embarked on a voyage to an area of the Atlantic seafloor where the mantle lies exposed without any crust covering, midway between the Cape Verde Islands and the Caribbean Sea. The exposed site lies approximately three kilometres beneath the ocean surface and covers thousands of square kilometres. A relatively difficult attempt to retrieve samples from the Earth's mantle was scheduled for later in 2007. The Chikyu Hakken mission attempted to use the Japanese vessel *Chikyū* to drill up to 7,000 m (23,000 ft) below the seabed. This is nearly three times as deep as preceding oceanic drillings.

A novel method of exploring the uppermost few hundred kilometres of the Earth was recently proposed, consisting of a small, dense, heat-generating probe which melts its way down through the crust and mantle while its position and progress are tracked by acoustic signals generated in the rocks. The probe consists of an outer sphere of tungsten about one metre in diameter with a cobalt-60 interior acting as a radioactive heat source. It was calculated that such a probe will reach the oceanic Moho in less than 6 months and attain minimum depths of well over 100 km (62 mi) in a few decades beneath both oceanic and continental lithosphere.

Exploration can also be aided through computer simulations of the evolution of the mantle. In 2009, a supercomputer application provided new insight into the distribution of mineral deposits, especially isotopes of iron, from when the mantle developed 4.5 billion years ago.

Lithosphere

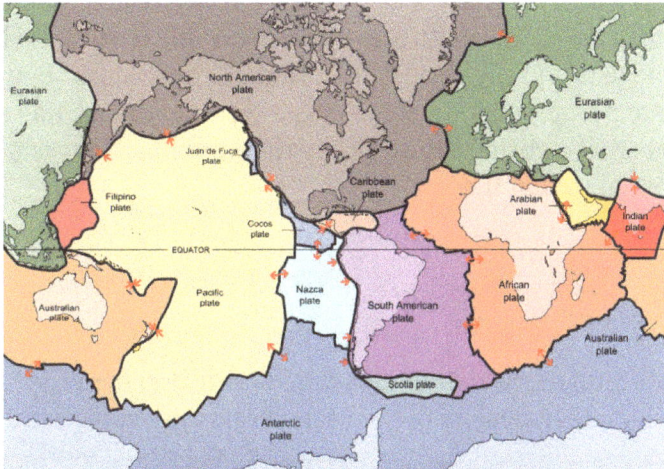

The tectonic plates of the lithosphere on Earth

A lithosphere is the rigid, outermost shell of a terrestrial-type planet or natural satellite that is defined by its rigid mechanical properties. On Earth, it is composed of the crust and the portion of the upper mantle that behaves elastically on time scales of thousands of years or greater. The outermost shell of a rocky planet, the crust, is defined on the basis of its chemistry and mineralogy.

Earth's Lithosphere

Earth's lithosphere includes the crust and the uppermost mantle, which constitute the hard and rigid outer layer of the Earth. The lithosphere is subdivided into tectonic plates. The uppermost part of the lithosphere that chemically reacts to the atmosphere, hydrosphere and biosphere through the soil forming process is called the pedosphere. The lithosphere is underlain by the asthenosphere which is the weaker, hotter, and deeper part of the upper mantle. The boundary between the lithosphere and the underlying asthenosphere is known as the Lithosphere-Asthenosphere boundary and is defined by a difference in response to stress: the lithosphere remains rigid for very long periods of geologic time in which it deforms elastically and through brittle failure, while the asthenosphere deforms viscously and accommodates strain through plastic deformation. The study of past and current formations of landscapes is called geomorphology.

History

The concept of the lithosphere as Earth's strong outer layer was described by A.E.H. Love in his 1911 monograph "Some problems of Geodynamics" and further developed by Joseph Barrell, who wrote a series of papers about the concept and introduced the term "lithosphere". The concept was based on the presence of significant gravity anomalies over continental crust, from which he inferred that there must exist a strong upper layer (which he called the lithosphere) above a weaker layer which could flow (which he called the asthenosphere). These ideas were expanded by Reginald Aldworth Daly in 1940 with his seminal work "Strength and Structure of the Earth" and have been broadly accepted by geologists and geophysicists. Although these ideas about lithosphere and asthenosphere were developed long before plate tectonic theory was articulated in the 1960s, the concepts that a strong lithosphere exists and that this rests on a weak asthenosphere are essential to that theory.

Types

There are two types of lithosphere:

- Oceanic lithosphere, which is associated with oceanic crust and exists in the ocean basins (mean density of about 2.9 grams per cubic centimeter)

- Continental lithosphere, which is associated with continental crust (mean density of about 2.7 grams per cubic centimeter)

The thickness of the lithosphere is considered to be the depth to the isotherm associated with the transition between brittle and viscous behavior. The temperature at which olivine begins to deform viscously (~1000 °C) is often used to set this isotherm because olivine is generally the weakest mineral in the upper mantle. Oceanic lithosphere is typically about 50–140 km thick (but beneath the mid-ocean ridges is no thicker than the crust), while continental lithosphere has a range in thickness from about 40 km to perhaps 280 km; the upper ~30 to ~50 km of typical continental lithosphere is crust. The mantle part of the lithosphere consists largely of peridotite. The crust is distinguished from the upper mantle by the change in chemical composition that takes place at the Moho discontinuity.

Oceanic Lithosphere

Oceanic lithosphere consists mainly of mafic crust and ultramafic mantle (peridotite) and is denser than continental lithosphere, for which the mantle is associated with crust made of felsic rocks. Oceanic lithosphere thickens as it ages and moves away from the mid-ocean ridge. This thickening occurs by conductive cooling, which converts hot asthenosphere into lithospheric mantle and causes the oceanic lithosphere to become increasingly thick and dense with age. The thickness of the mantle part of the oceanic lithosphere can be approximated as a thermal boundary layer that thickens as the square root of time.

$$h \sim 2\sqrt{\kappa t}$$

Here, h is the thickness of the oceanic mantle lithosphere, κ is the thermal diffusivity (approximately 10^{-6} m²/s) for silicate rocks, and t is the age of the given part of the lithosphere. The age is often equal to L/V, where L is the distance from the spreading centre of mid-oceanic ridge, and V is velocity of the lithospheric plate.

Oceanic lithosphere is less dense than asthenosphere for a few tens of millions of years but after this becomes increasingly denser than asthenosphere. This is because the chemically differentiated oceanic crust is lighter than asthenosphere, but thermal contraction of the mantle lithosphere makes it more dense than the asthenosphere. The gravitational instability of mature oceanic lithosphere has the effect that at subduction zones, oceanic lithosphere invariably sinks underneath the overriding lithosphere, which can be oceanic or continental. New oceanic lithosphere is constantly being produced at mid-ocean ridges and is recycled back to the mantle at subduction zones. As a result, oceanic lithosphere is much younger than continental lithosphere: the oldest oceanic lithosphere is about 170 million years old, while parts of the continental lithosphere are billions of years old. The oldest parts of continental lithosphere underlie cratons, and the mantle lithosphere there is thicker and less dense than typical; the relatively low density of such mantle "roots of cratons" helps to stabilize these regions.

Subducted Lithosphere

Geophysical studies in the early 21st century posit that large pieces of the lithosphere have been subducted into the mantle as deep as 2900 km to near the core-mantle boundary, while others "float" in the upper mantle, while some stick down into the mantle as far as 400 km but remain "attached" to the continental plate above, similar to the extent of the "tectosphere" proposed by Jordan in 1988.

Mantle Xenoliths

Geoscientists can directly study the nature of the subcontinental mantle by examining mantle xenoliths brought up in kimberlite, lamproite, and other volcanic pipes. The histories of these xenoliths have been investigated by many methods, including analyses of abundances of isotopes of osmium and rhenium. Such studies have confirmed that mantle lithospheres below some cratons have persisted for periods in excess of 3 billion years, despite the mantle flow that accompanies plate tectonics.

References

- Condie, Kent C. (1997). Plate tectonics and crustal evolution. Butterworth-Heinemann. p. 123. ISBN 978-0-7506-3386-4. Retrieved 21 May 2010.

- Kearey, Philip; Vine, Frederick J. (1996). Global Tectonics (2 ed.). Wiley-Blackwell. pp. 41–42. ISBN 978-0-86542-924-6. Retrieved 21 May 2010.

- Burns, Roger George (1993). Mineralogical Applications of Crystal Field Theory. Cambridge University Press. p. 354. ISBN 0-521-43077-1. Retrieved 2007-12-26.

- Anderson, Don L. (2007) New Theory of the Earth. Cambridge University Press. ISBN 978-0-521-84959-3, ISBN 0-521-84959-4

- Jackson, Ian (1998). The Earth's Mantle - Composition, Structure, and Evolution. Cambridge University Press. pp. 311–378. ISBN 0-521-78566-9.

- Skinner, B.J. & Porter, S.C.: Physical Geology, page 17, chapt. The Earth: Inside and Out, 1987, John Wiley & Sons, ISBN 0-471-05668-5

- Rudoph, Maxwell (11 December 2015). "Viscosity jump in Earth's mid-mantle". Science. Retrieved 16 January 2016.

- Sumner, Thomas (10 December 2015). "Gooey rock in mantle thickens 1,000 kilometers down". Science News. Retrieved 16 January 2016.

- Neumann, W.; et al. (2014). "Differentiation of Vesta: Implications for a shallow magma ocean". Earth and Planetary Science Letters. 395: 267–280. doi:10.1016/j.epsl.2014.03.033.

- "Rare Diamond confirms that Earth's mantle holds an ocean's worth of water". Scientific American. March 12, 2014. Retrieved March 13, 2014.

- Pasyanos M. E. (2008-05-15). "Lithospheric Thickness Modeled from Long Period Surface Wave Dispersion" (PDF). Retrieved 2014-04-25.

Key Concepts of Plate Tectonics

Continental drifting is the shifting of the continents of the Earth and it has been occurring over a period of millions of years. The other key concepts of plate tectonics that have been explained in this chapter are divergent boundary, continental collision, plate reconstruction, crustal recycling etc. The section strategically encompasses and incorporates the major components and key concepts of plate tectonics, providing a complete understanding.

Mantle Convection

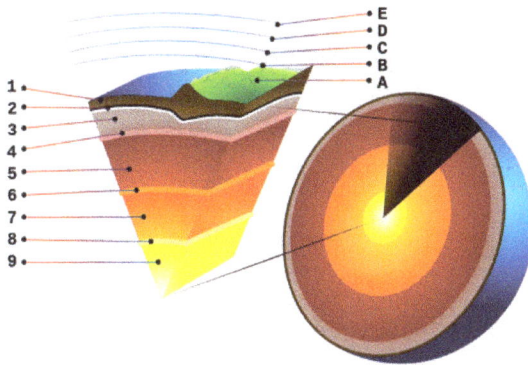

Earth cross-section showing location of upper (3) and lower (5) mantle

Mantle convection is the slow creeping motion of Earth's solid silicate mantle caused by convection currents carrying heat from the interior of the Earth to the surface. The Earth's surface lithosphere, which rides atop the asthenosphere (the two components of the upper mantle), is divided into a number of plates that are continuously being created and consumed at their opposite plate boundaries. Accretion occurs as mantle is added to the growing edges of a plate, associated with seafloor spreading. This hot added material cools down by conduction and convection of heat. At the consumption edges of the plate, the material has thermally contracted to become dense, and it sinks under its own weight in the process of subduction usually at an ocean trench.

This *subducted* material sinks through the Earth's interior. Some subducted material appears to reach the lower mantle, while in other regions, this material is impeded from sinking further, possibly due to a phase transition from spinel to silicate perovskite and magnesiowustite, an endothermic reaction.

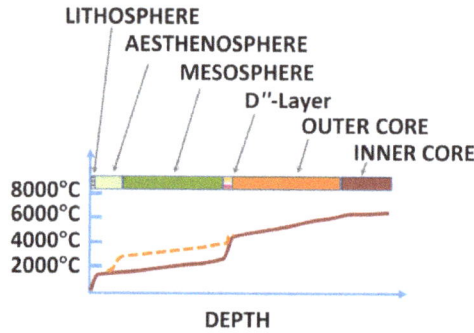

Calculated Earth's temperature vs. depth. Dashed curve: Layered mantle convection; Solid curve: Whole mantle convection.

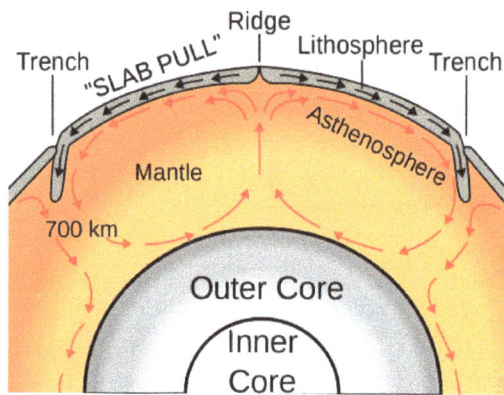

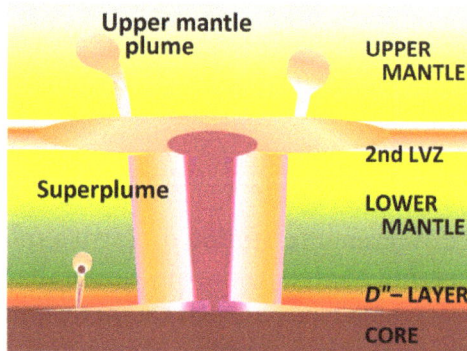

Whole mantle convection

A superplume generated by cooling processes in the mantle.

The subducted oceanic crust triggers volcanism, although the basic mechanisms are varied. Volcanism may occur due to processes that add buoyancy to partially melted mantle causing an upward flow due to a decrease in density of the partial melt.

Secondary forms of convection that may result in surface volcanism are postulated to occur as a consequence of intraplate extension and mantle plumes.

It is because the mantle can convect that the tectonic plates are able to move around the Earth's surface.

Mantle convection seems to have been much more active during the Hadean period, resulting in gravitational sorting of heavier molten iron, and nickel elements and sulphides in the core, and lighter silicate minerals in the mantle.

Types of Convection

During the late 20th century, there was significant debate within the geophysics community as to whether convection is likely to be 'layered' or 'whole'. Although elements of this debate still continue, results from seismic tomography, numerical simulations of mantle convection and examination of Earth's gravitational field are all beginning to suggest the existence of 'whole' mantle convection, at least at the present time. In this model, cold, subducting oceanic lithosphere descends all the way from the surface to the core-mantle boundary (CMB) and hot plumes rise from the CMB all the way to the surface. This picture is strongly based on the results of global seismic tomography models, which typically show slab and plume-like anomalies crossing the mantle transition zone.

Although it is now well accepted that subducting slabs cross the mantle transition zone and descend into the lower mantle, debate about the existence and continuity of plumes persists, with important implications for the style of mantle convection. This debate is linked to the controversy regarding whether intraplate volcanism is caused by shallow, upper-mantle processes or by plumes from the lower mantle. Many geochemistry studies have argued that the lavas erupted in intraplate areas are different in composition from shallow-derived mid ocean ridge basalts (MORB). Specifically, they typically have elevated Helium-3 - Helium-4 ratios. Being a primordial nuclide, Helium-3 is not naturally produced on earth. It also quickly escapes from earth's atmosphere when erupted. The elevated He-3/He-4 ratio of Ocean Island Basalts (OIBs) suggest that they must be sources from a part of the earth that has not previously been melted and reprocessed in the same way as MORB source has been. This has been interpreted as their originating from a different, less well-mixed, region, suggested to be the lower mantle. Others, however, have pointed out that geochemical differences could indicate the inclusion of a small component of near-surface material from the lithosphere.

Speed of Convection

Typical mantle convection speed is 20 mm/yr near the crust but can vary quite a bit. The small scale convection in the upper mantle is much faster than the convection near the core. A single shallow convection cycle takes on the order of 50 million years, though deeper convection can be closer to 200 million years.

Creep in the Mantle

Since the mantle is primarily composed of olivine $((Mg,Fe)_2SiO_4)$, the rheological characteristics of the mantle are largely those of olivine. Additionally, due to the varying temperatures and pressures between the lower and upper mantle, a variety of creep processes can occur with dislocation creep dominating in the lower mantle and diffusional creep occasionally dominating in the upper mantle. However, there is a large transition region in creep processes between the upper and lower mantle and even within each section, creep properties can change strongly with location and thus temperature and pressure. In the power law creep regions, the creep equation fitted to data with n = 3-4 is standard.

The strength of olivine not only scales with its melting temperature, but also is very sensitive to water and silica content. The solidus depression by impurities, primarily Ca, Al, and Na, and pressure affects creep behavior and thus contributes to the change in creep mechanisms with location. While creep behavior is generally plotted as homologous temperature versus stress, in the case of the mantle it is often more useful to look at the pressure dependence of stress. Though stress is simple force over area, defining the area is difficult in geology. Equation 1 demonstrates the pressure dependence of stress. Since it is very difficult to simulate the high pressures in the mantle (1MPa at 300–400 km), the low pressure laboratory data is usually extrapolated to high pressures by applying creep concepts from metallurgy.

$$(1)\ \left(\frac{\partial \ln \sigma}{\partial P}\right)_{T,\dot{\varepsilon}} = \left(\frac{1}{TT_m}\right) * \left(\frac{\partial \ln \sigma}{\partial (1/T)}\right)_{P,\dot{\varepsilon}} * \frac{dT_m}{dP}$$

Most of the mantle has homologous temperatures of 0.65-0.75 and experiences strain rates of $10^{-14} - 10^{-16}$ 1/s. Stresses in mantle are dependent on density, gravity, thermal expansion coefficients, temperature differences driving convection, and distance convection occurs over, all of which give stresses around a fraction of 3-30MPa. Due to the large grain sizes (at low stresses as high as several mm), it is unlikely that Nabarro-Herring (NH) creep truly dominates. Given the large grain sizes, dislocation creep tends to dominate. 14 MPa is the stress below which diffusional creep dominates and above which power law creep dominates at 0.5Tm of olivine. Thus, even for relatively low temperatures, the stress diffusional creep would operate at is too low for realistic conditions. Though the power law creep rate increases with increasing water content due to weakening, reducing activation energy of diffusion and thus increasing the NH creep rate, NH is generally still not large enough to dominate. Nevertheless, diffusional creep can dominate in very cold or deep parts of the upper mantle. Additional deformation in the mantle can be attributed to transformation enhanced ductility. Below 400 km, the olivine undergoes a pressure induced phase transformation into spinel and can cause more deformation due to the increased ductility. Further evidence for the dominance of power law creep comes from preferred lattice orientations as a result of deformation. Under dislocation creep, crystal structures reorient into lower stress orientations. This does not happen under diffusional creep, thus observation of preferred orientations in samples lends credence to the dominance of dislocation creep.

Continental Drift

The continental drift of the last 150 million years

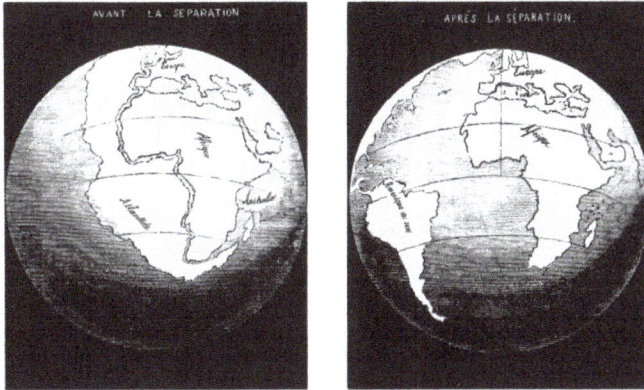

Antonio Snider-Pellegrini's Illustration of the closed and opened Atlantic Ocean (1858).

Continental drift is the movement of the Earth's continents relative to each other, thus appearing to "drift" across the ocean bed. The speculation that continents might have 'drifted' was first put forward by Abraham Ortelius in 1596. The concept was independently and more fully developed by Alfred Wegener in 1912, but his theory was rejected by some for lack of a mechanism (though this was supplied later by Arthur Holmes) and others because of prior theoretical commitments. The idea of continental drift has been subsumed by the theory of plate tectonics, which explains how the continents move.

In 1858 Antonio Snider-Pellegrini created two maps demonstrating how the American and African continents might have once fit together.

History

Early History

Abraham Ortelius (Ortelius 1596), Theodor Christoph Lilienthal (1756), Alexander von

Humboldt (1801 and 1845), Antonio Snider-Pellegrini (Snider-Pellegrini 1858), and others had noted earlier that the shapes of continents on opposite sides of the Atlantic Ocean (most notably, Africa and South America) seem to fit together. W. J. Kious described Ortelius' thoughts in this way:

Abraham Ortelius in his work Thesaurus Geographicus ... suggested that the Americas were "torn away from Europe and Africa ... by earthquakes and floods" and went on to say: "The vestiges of the rupture reveal themselves, if someone brings forward a map of the world and considers carefully the coasts of the three [continents]."

Writing in 1889, Alfred Russel Wallace remarks "It was formerly a very general belief, even amongst geologists, that the great features of the earth's surface, no less than the smaller ones, were subject to continual mutations, and that during the course of known geological time the continents and great oceans had again and again changed places with each other." He quotes Charles Lyell as saying "Continents, therefore, although permanent for whole geological epochs, shift their positions entirely in the course of ages" and claims that the first to throw doubt on this was James Dwight Dana in 1849.

In his *Manual of Geology*, 1863, Dana says "The continents and oceans had their general outline or form defined in earliest time. This has been proved with respect to North America from the position and distribution of the first beds of the Silurian - those of the Potsdam epoch. ... and this will probably prove to the case in Primordial time with the other continents also". Dana was enormously influential in America - his *Manual of Mineralogy* is still in print in revised form - and the theory became known as *Permanence theory*.

This appeared to be confirmed by the exploration of the deep sea beds conducted by the Challenger expedition, 1872-6, which showed that contrary to expectation, land debris brought down by rivers to the ocean is deposited comparatively close to the shore in what is now known as the continental shelf. This suggested that the oceans were a permanent feature of the earth's surface, and did not change places with the continents.

Wegener and his Predecessors

Alfred Wegener

The speculation that the American continents had once formed a single landmass with Europe and Asia before assuming the present shapes and positions was suggested by several scientists before Alfred Wegener's 1912 paper. Although Wegener's theory was formed independently and was more complete than those of his predecessors, Wegener later credited a number of past authors with similar ideas: Franklin Coxworthy (between 1848 and 1890), Roberto Mantovani (between 1889 and 1909), William Henry Pickering (1907) and Frank Bursley Taylor (1908). In addition, Eduard Suess had proposed a supercontinent Gondwana in 1885 and the Tethys Ocean in 1893, assuming a land-bridge between the present continents submerged in the form of a geosyncline, and John Perry had written an 1895 paper proposing that the earth's interior was fluid, and disagreeing with Lord Kelvin on the age of the earth.

For example: the similarity of southern continent geological formations had led Roberto Mantovani to conjecture in 1889 and 1909 that all the continents had once been joined into a supercontinent; Wegener noted the similarity of Mantovani's and his own maps of the former positions of the southern continents. In Mantovani's conjecture, through volcanic activity due to thermal expansion this continent broke and the new continents drifted away from each other because of further expansion of the rip-zones, where the oceans now lie. This led Mantovani to propose an Expanding Earth theory which has since been shown to be incorrect.

Continental drift without expansion was proposed by Frank Bursley Taylor, who suggested in 1908 (published in 1910) that the continents were moved into their present positions by a process of "continental creep". In a later paper he proposed that this occurred by their being dragged towards the equator by tidal forces during the hypothesized capture of the moon in the Cretaceous, resulting in "general crustal creep" toward the equator. Although his proposed mechanism was wrong, he was the first to realize the insight that one of the effects of continental motion would be the formation of mountains, and attributed the formation of the Himalayas to the collision between the Indian subcontinent with Asia. Wegener said that of all those theories, Taylor's, although not fully developed, had the most similarities to his own. In the mid-20th century, the theory of continental drift was referred to as the "Taylor-Wegener hypothesis", although this terminology eventually fell out of common use.

Alfred Wegener first presented his hypothesis to the German Geological Society on January 6, 1912. His hypothesis was that the continents had once formed a single landmass, called Pangea, before breaking apart and drifting to their present locations.

Wegener was the first to use the phrase "continental drift" (1912, 1915) (in German "die Verschiebung der Kontinente" – translated into English in 1922) and formally publish the hypothesis that the continents had somehow "drifted" apart. Although he presented much evidence for continental drift, he was unable to provide a convincing explanation for the physical processes which might have caused this drift. His suggestion that the continents had been pulled apart by the centrifugal pseudoforce (*Polflucht*) of the

Earth's rotation or by a small component of astronomical precession was rejected as calculations showed that the force was not sufficient. The Polflucht hypothesis was also studied by Paul Sophus Epstein in 1920 and found to be implausible.

Rejection of Wegener's Theory, 1910s–1950s

The theory of continental drift was not accepted for many years. One problem was that a plausible driving force was missing. A second problem was that Wegener's estimate of the velocity of continental motion, 250 cm/year, was implausibly high. (The currently accepted rate for the separation of the Americas from Europe and Africa is about 2.5 cm/year). And it did not help that Wegener was not a geologist. Other geologists also believed that the evidence that Wegener had provided was not sufficient. It is now accepted that the plates carrying the continents do move across the Earth's surface, although not as fast as Wegener believed; ironically one of the chief outstanding questions is the one Wegener failed to resolve: what is the nature of the forces propelling the plates?

The British geologist Arthur Holmes championed the theory of continental drift at a time when it was deeply unfashionable. He proposed in 1931 that the Earth's mantle contained convection cells that dissipated radioactive heat and moved the crust at the surface. His *Principles of Physical Geology*, ending with a chapter on continental drift, was published in 1944.

David Attenborough, who attended university in the second half of the 1940s, recounted an incident illustrating its lack of acceptance then: "I once asked one of my lecturers why he was not talking to us about continental drift and I was told, sneeringly, that if I could prove there was a force that could move continents, then he might think about it. The idea was moonshine, I was informed."

Geological maps of the time showed huge land bridges spanning the Atlantic and Indian oceans to account for the similarities of fauna and flora and the divisions of the Asian continent in the Permian era but failing to account for glaciation in India, Australia and South Africa.

As late as 1953 – just five years before Carey introduced the theory of plate tectonics – the theory of continental drift was rejected by the physicist Scheidegger on the following grounds.

- First, it had been shown that floating masses on a rotating geoid would collect at the equator, and stay there. This would explain one, but only one, mountain building episode between any pair of continents; it failed to account for earlier orogenic episodes.

- Second, masses floating freely in a fluid substratum, like icebergs in the ocean, should be in isostatic equilibrium (in which the forces of gravity and buoyancy are in balance). But gravitational measurements showed that many areas are not in isostatic equilibrium.

- Third, there was the problem of why some parts of the Earth's surface (crust) should have solidified while other parts were still fluid. Various attempts to explain this foundered on other difficulties.

Geophysicist Jack Oliver is credited with providing seismologic evidence supporting plate tectonics which encompassed and superseded continental drift with the article "Seismology and the New Global Tectonics", published in 1968, using data collected from seismologic stations, including those he set up in the South Pacific.

It is now known that there are two kinds of crust: continental crust and oceanic crust. Continental crust is inherently lighter and its composition is different from oceanic crust, but both kinds reside above a much deeper "plastic" mantle. Oceanic crust is created at spreading centers, and this, along with subduction, drives the system of plates in a chaotic manner, resulting in continuous orogeny and areas of isostatic imbalance. The theory of plate tectonics explains all this, including the movement of the continents, better than Wegener's theory.

Evidence of Continental Drift

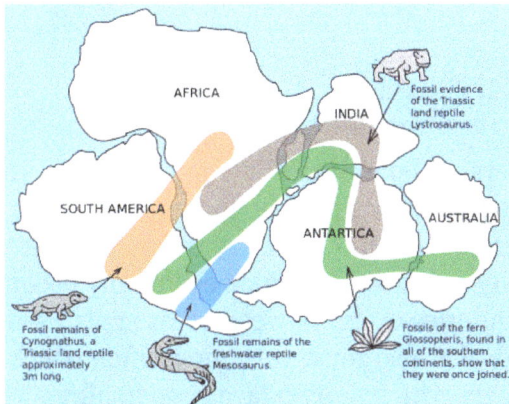

Fossil patterns across continents (Gondwanaland).

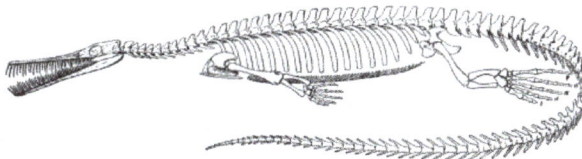

Mesosaurus skeleton, MacGregor, 1908.

Evidence for the movement of continents on tectonic plates is now extensive. Similar plant and animal fossils are found around the shores of different continents, suggesting that they were once joined. The fossils of *Mesosaurus*, a freshwater reptile rather like a small crocodile, found both in Brazil and South Africa, are one example; another is the discovery of fossils of the land reptile *Lystrosaurus* in rocks of the same age at locations in Africa, India, and Antarctica. There is also living evidence—the same animals being found on two continents. Some earthworm families (e.g. Ocnerodrilidae, Acanthodrilidae, Octochaetidae) are found in South America and Africa, for instance.

The complementary arrangement of the facing sides of South America and Africa is obvious, but is a temporary coincidence. In millions of years, slab pull and ridge-push, and other forces of tectonophysics, will further separate and rotate those two continents. It was this temporary feature which inspired Wegener to study what he defined as continental drift, although he did not live to see his hypothesis generally accepted.

Widespread distribution of Permo-Carboniferous glacial sediments in South America, Africa, Madagascar, Arabia, India, Antarctica and Australia was one of the major pieces of evidence for the theory of continental drift. The continuity of glaciers, inferred from oriented glacial striations and deposits called tillites, suggested the existence of the supercontinent of Gondwana, which became a central element of the concept of continental drift. Striations indicated glacial flow away from the equator and toward the poles, based on continents' current positions and orientations, and supported the idea that the southern continents had previously been in dramatically different locations, as well as being contiguous with each other.

Divergent Boundary

In plate tectonics, a divergent boundary or divergent plate boundary (also known as a constructive boundary or an extensional boundary) is a linear feature that exists between two tectonic plates that are moving away from each other. Divergent boundaries within continents initially produce rifts which eventually become rift valleys. Most active divergent plate boundaries occur between oceanic plates and exist as mid-oceanic ridges. Divergent boundaries also form volcanic islands which occur when the plates move apart to produce gaps which molten lava rises to fill.

Current research indicates that complex convection within the Earth's mantle allows material to rise to the base of the lithosphere beneath each divergent plate boundary. This supplies the area with vast amounts of heat and a reduction in pressure that melts rock from the asthenosphere (or upper mantle) beneath the rift area forming large flood basalt or lava flows. Each eruption occurs in only a part of the plate boundary at any one time, but when it does occur, it fills in the opening gap as the two opposing plates move away from each other.

Over millions of years, tectonic plates may move many hundreds of kilometers away from both sides of a divergent plate boundary. Because of this, rocks closest to a boundary are younger than rocks further away on the same plate.

Description

Bridge across the Álfagjá rift valley in southwest Iceland, that is part of the boundary between the Eurasian and North American continental tectonic plates.

At divergent boundaries, two plates move apart from each other and the space that this creates is filled with new crustal material sourced from molten magma that forms below. The origin of new divergent boundaries at triple junctions is sometimes thought to be associated with the phenomenon known as hotspots. Here, exceedingly large convective cells bring very large quantities of hot asthenospheric material near the surface and the kinetic energy is thought to be sufficient to break apart the lithosphere. The hot spot which may have initiated the Mid-Atlantic Ridge system currently underlies Iceland which is widening at a rate of a few centimeters per year.

Divergent boundaries are typified in the oceanic lithosphere by the rifts of the oceanic ridge system, including the Mid-Atlantic Ridge and the East Pacific Rise, and in the continental lithosphere by rift valleys such as the famous East African Great Rift Valley. Divergent boundaries can create massive fault zones in the oceanic ridge system. Spreading is generally not uniform, so where spreading rates of adjacent ridge blocks are different, massive transform faults occur. These are the fracture zones, many bearing names, that are a major source of submarine earthquakes. A sea floor map will show a rather strange pattern of blocky structures that are separated by linear features perpendicular to the ridge axis. If one views the sea floor between the fracture zones as conveyor belts carrying the ridge on each side of the rift away from the spreading center the action becomes clear. Crest depths of the old ridges, parallel to the current spreading center, will be older and deeper... (from thermal contraction and subsidence).

It is at mid-ocean ridges that one of the key pieces of evidence forcing acceptance of the seafloor spreading hypothesis was found. Airborne geomagnetic surveys showed a strange pattern of symmetrical magnetic reversals on opposite sides of ridge centers. The pattern was far too regular to be coincidental as the widths of the opposing bands were too closely matched. Scientists had been studying polar reversals and the link was made by Lawrence W. Morley, Frederick John Vine and Drummond Hoyle Matthews in the Morley–Vine–Matthews hypothesis. The magnetic banding directly corresponds with the Earth's polar reversals. This was confirmed by measuring the ages of the rocks within each band. The banding furnishes a map in time and space of both spreading rate and polar reversals.

Examples

- Mid-Atlantic Ridge

- Red Sea Rift

- Baikal Rift Zone

- East African Rift

- East Pacific Rise

- Gakkel Ridge

- Galapagos Rise

- Explorer Ridge

- Juan de Fuca Ridge

- Pacific-Antarctic Ridge

- West Antarctic Rift

- Great Rift Valley

Other Plate Boundary Types

- Convergent boundary

- Transform boundary

Continental Collision

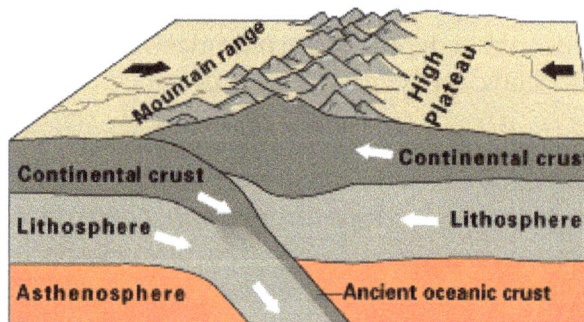

Continental-continental convergence
Cartoon of a tectonic collision between two continents

Continental collision is a phenomenon of the plate tectonics of Earth that occurs at convergent boundaries. Continental collision is a variation on the fundamental process of subduction, whereby the subduction zone is destroyed, mountains produced, and two continents sutured together. Continental collision is known only to occur on Earth.

Continental collision is not an instantaneous event, but may take several tens of millions of years before the faulting and folding caused by collisions stops. The collision between India and Asia has been going on for about 50 million years already and shows no signs of abating. Collision between East and West Gondwana to form the East African Orogen took about 100 million years from beginning (610 Ma) to end (510 Ma). Collision between Gondwana and Laurasia to form Pangea occurred in a relatively brief interval, about 50 million years long.

Subduction Zone: the Collision Site

The process begins as two continents (different bits of continental crust), separated across a tract of ocean (and oceanic crust), approach each other, while the oceanic crust is slowly consumed at a subduction zone. The subduction zone runs along the edge of one of the continents and dips under it, raising volcanic mountain chains at some distance behind it, such as the Andes of South America today. Subduction involves the whole lithosphere, the density of which is largely controlled by the nature of the crust it carries. Oceanic crust is thin (~6 km thick) and dense (about 3.3 g/cm³), consisting of basalt, gabbro, and peridotite. Consequently, most oceanic crust is subducted easily at an oceanic trench. In contrast, continental crust is thick (~45 km thick) and buoyant, composed mostly of granitic rocks (average density about 2.5 g/cm³). Continental crust is subducted with difficulty, but is subducted to depths of 90-150 km or more, as evidenced by ultra-high pressure (UHP) metamorphic suites. Normal subduction continues as long as the ocean exists, but the subduction system is disrupted as the continent carried by the downgoing plate enters the trench. Because it contains thick continental crust, this lithosphere is less dense than the underlying asthenospheric mantle and normal subduction is disrupted. The volcanic arc on the upper plate is slowly extinguished. Resisting subduction, the crust buckles up and under, raising mountains where a trench used to be. The position of the trench becomes a zone that marks the suture between the two continental terranes. Suture zones are often marked by fragments of the pre-existing oceanic crust and mantle rocks, known as ophiolites.

Deep Subduction of Continental Crust

The continental crust on the downgoing plate is deeply subducted as part of the downgoing plate during collision, defined as buoyant crust entering a subduction zone. An unknown proportion of subducted continental crust returns to the surface as ultra-high pressure (UHP) metamorphic terranes, which contain metamorphic coesite and/or diamond plus or minus unusual silicon-rich garnets and/or potassium-bearing pyroxenes. The presence of these minerals demonstrate subduction of continental crust to at least 90–140 km deep. Examples of UHP terranes are known from the Dabie–Sulu belt of east-central China, the

Western Alps, the Himalaya of India, the Kokchetav Massif of Kazakhstan, the Bohemian Massif of Europe, the North Qaidam of Northwestern China, the Western Gneiss Region of Norway, and Mali. Most UHP terranes consist of an imbricated sheets or nappes. The fact that most UHP terranes consist of thin sheets suggests that much thicker, volumetrically dominant tracts of continental crust are more deeply subducted.

Orogeny and Collapse

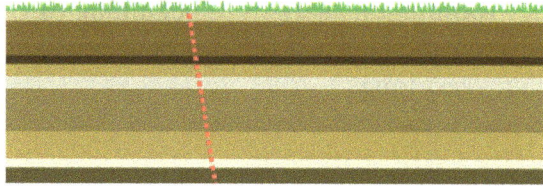

Mountain formation by a reverse fault movement

An orogeny is underway when mountains begin to grow in the collision zone. There are other modes of mountain formation and orogeny but certainly continental collision is one of the most important. Rainfall and snowfall increase on the mountains as these rise, perhaps at a rate of a few millimeters per year (at a growth rate of 1 mm/year, a 5,000 m tall mountain can form in 5 million years, a time period that is less than 10% of the life of a typical collision zone). River systems form, and glaciers may grow on the highest peaks. Erosion accelerates as the mountains rise, and great volumes of sediment are shed into the rivers, which carry sediment away from the mountains to be deposited in sedimentary basins in the surrounding lowlands. Crustal rocks are thrust faulted over the sediments and the mountain belt broadens as it rises in height. A crustal root also develops, as required by isostasy; mountains can be high if underlain by thicker crust. Crustal thickening may happen as a result of crustal shortening or when one crust overthrusts the other. Thickening is accompanied by heating, so the crust becomes weaker as it thickens. The lower crust begins to flow and collapse under the growing mountain mass, forming rifts near the crest of the mountain range. The lower crust may partially melt, forming anatectic granites which then rise into the overlying units, forming granite intrusions. Crustal thickening provides one of two negative feedbacks on mountain growth in collision zones, the other being erosion. The popular notion that erosion is responsible for destroying mountains is only half correct - viscous flow of weak lower mantle also reduces relief with time, especially once the collision is complete and the two continents are completely sutured. Convergence between the continents continues because the crust is still being pulled down by oceanic lithosphere sinking in the subduction zone to either side of the collision as well as beneath the impinging continent.

The pace of mountain building associated with the collision is measured by radiometric dating of igneous rocks or units that have been metamorphosed during the collision and by examining the record of sediments shed from the rising mountains into the surrounding basins. The pace of ancient convergence can be determined with paleomagnetic measurements, while the present rate of convergence can be measured with GPS.

Far-field Effects

The effects of the collision are felt far beyond the immediate site of collision and mountain-building. As convergence between the two continents continues, the region of crustal thickening and elevation will become broader. If there is an oceanic free face, the adjacent crustal blocks may move towards it. As an example of this, the collision of India with Asia forced large regions of crust to move south to form modern Southeast Asia. Another example is the collision of Arabia with Asia, which is squeezing the Anatolian Plate (present day Turkey). As a result, Turkey is moving west and south into the Mediterranean Sea and away from the collision zone. These far-field effects may result in the formation of rifts, and rift valleys such as that occupied by Lake Baikal, the deepest lake on Earth.

Fossil Collision Zones

Continental collisions are a critical part of the Supercontinent cycle and have happened many times in the past. Ancient collision zones are deeply eroded but may still be recognized because these mark sites of intense deformation, metamorphism, and plutonic activity that separate tracts of continental crust having different geologic histories prior to the collision. Old collision zones are commonly called "suture zones" by geologists, because this is where two previous continents are joined or *sutured* together.

Plate Reconstruction

Plate reconstruction is the process of reconstructing the positions of tectonic plates relative to each other (relative motion) or to other reference frames, such as the earth's magnetic field or groups of hotspots, in the geological past. This helps determine the shape and make-up of ancient supercontinents and provides a basis for paleogeographic reconstructions.

Defining Plate Boundaries

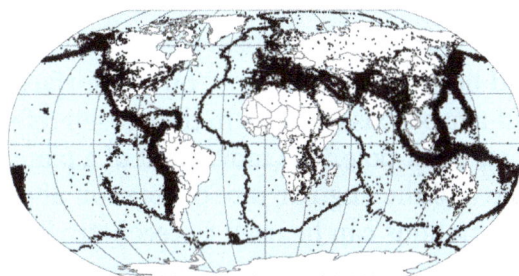

Earthquake epicenters 1963–98

An important part of reconstructing past plate configurations is to define the edges of areas of the lithosphere that have acted independently at some time in the past.

Present Plate Boundaries

Most present plate boundaries are easily identifiable from the pattern of recent seismicity. This is now backed up by the use of GPS data, to confirm the presence of significant relative movement between plates.

Past Plate Boundaries

Identifying past (but now inactive) plate boundaries within current plates is generally based on evidence for an ocean that has now closed up. The line where the ocean used to be is normally marked by pieces of the crust from that ocean, included in the collision zone, known as ophiolites. The line across which two plates became joined to form a single larger plate, is known as a suture.

In many orogenic belts, the collision is not just between two plates, but involves the sequential accretion of smaller terranes. Terranes are smaller pieces of continental crust that have been caught up in an orogeny, such as continental fragments or island arcs.

Reference Frames

Plate motions, both those observable now and in the past, are referred ideally to a reference frame that allows other plate motions to be calculated. For example, a central plate, such as the African plate, may have the motions of adjacent plates referred to it. By composition of reconstructions, additional plates can be reconstructed to the central plate. In turn, the reference plate may be reconstructed, together with the other plates, to another reference frame, such as the earth's magnetic field, as determined from paleomagnetic measurements of rocks of known age. A global hotspot reference frame has been postulated but there is now evidence that not all hotspots are necessarily fixed in their locations relative to one another or the earth's spin axis. However, there are groups of such hotspots that appear to be fixed within the constraints of available data, within particular mesoplates.

Euler Poles

The movement of a rigid body, such as a plate, on the surface of a sphere can be described as rotation about a fixed axis (relative to the chosen reference frame). This pole of rotation is known as an Euler pole. The movement of a plate is completely specified in terms of its Euler pole and the angular rate of rotation about the pole. Euler poles defined for current plate motions can be used to reconstruct plates in the recent past (few million years). At earlier stages of earth's history, new Euler poles need to be defined. It was suggested recently that paleomagnetic Euler poles derived from apparent polar wander path could potentially extend the upper age limit of marine geophysical observations and hot spot traces.

Estimating Past Plate Motions

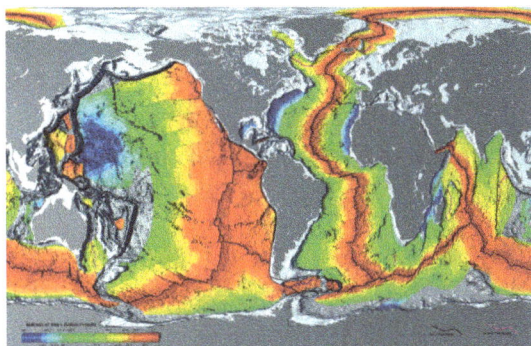

Ages of oceanic lithosphere

In order to move plates backward in time it is necessary to provide information on either relative or absolute positions of the plates being reconstructed such that an Euler pole can be calculated. These are quantitative methods of reconstruction.

Geometric Matching of Continental Borders

Certain fits between continents, particularly that between South America and Africa, were known long before the development of a theory that could adequately explain them. The reconstruction before Atlantic rifting by Bullard based on a least-squares fitting at the 500 fathom contour still provides the best match to paleomagnetic pole data for the two sides from the middle of Paleozoic to Late Triassic.

Plate Motion from Magnetic Stripes

Plate reconstructions in the recent geological past mainly use the pattern of magnetic stripes in oceanic crust to remove the effects of seafloor spreading. The individual stripes are dated from magnetostratigraphy so that their time of formation is known. Each stripe (and its mirror image) represents a plate boundary at a particular time in the past, allowing the two plates to be repositioned relative to one another. The oldest oceanic crust is of Jurassic age, providing a lower age limit of about 175 Ma for the use of such data. Reconstructions derived in this way are only relative.

Paleomagnetic Pole Data

Sampling

Paleomagnetic pole data is obtained by taking oriented samples of rock. Good quality poles have been recovered from many different rock types. In igneous rocks, the magnetic minerals have cooled through the curie point, typically providing good quality poles. Metamorphic rocks may contain new magnetic minerals but are not normally used to obtain poles due to their typically high magnetic anisotropy. Sedimentary rocks generally have a complex set of magnetic components such as derived clasts and mineral grains with their own inherited magnetization, primary magnetization acquired at the time of sedimentation and secondary magnetization related to subsequent diagenesis and authigenic mineral growth. Progressive demagnetization is used to discriminate between the components to identify the primary magnetization. The poles shown by the primary magnetization represent a record of their position relative to the magnetic pole at the time when the rock was formed. Suitable rocks of the right age are not present everywhere, so some pieces of the puzzle remain very unconstrained.

Good quality magnetic poles are added to the *Global Paleomagnetic Database*, which is accessible from the World Data Center A in the USA at Boulder, Colorado; and from the Norwegian Geological Survey in Trondheim.

Apparent Polar Wander Paths

Poles from different ages in a single area can be used to construct an apparent polar wander (APW) path. If paths from adjacent crustal fragments are identical, this is taken to indicate that there has been no relative movement between them during the period covered by the path. Divergence of APW paths indicates that the areas in question have acted independently in the past with the point of divergence marking the time at which they became joined.

Observed paleomagnetic poles show only the approximate latitude at the time but provide no constraint on the longitude. For this reason older plate reconstructions have much larger uncertainties associated with them. Paleomagnetic Euler rotations (rotation poles and angles) and the mantle structures, however, could provide clues to this longstanding challenge.

Hotspot Tracks

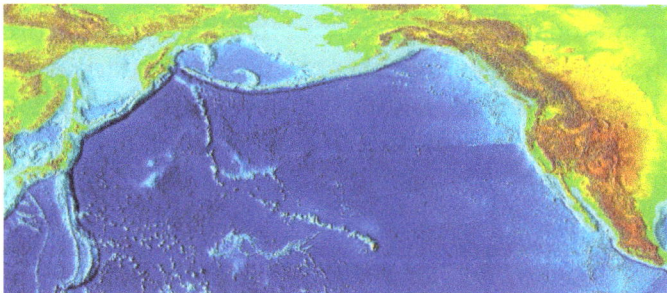

The Hawaiian-Emperor seamount chain

The presence of chains of volcanic islands and seamounts interpreted to have formed from fixed hotspots allows the plate on which they sit to be progressively restored so that a seamount is moved back over the hotspot at its time of formation. This method can be used back to the Early Cretaceous, the age of the oldest evidence for hotspot activity. This method gives an absolute reconstruction of both latitude and longitude, although before about 90 Ma there is evidence of relative motion between hotspot groups.

Slab Constraints

Once oceanic plates subduct in the lower mantle (slabs), they are assumed to sink in a near-vertical manner. With the help of seismic wave tomography, this can be used to constrain plate reconstructions at first order back to the Permian.

Apparent Polar Wander Paths Geometric Parameterizations

Paleomagnetic Euler poles derived by geometrizing apparent polar wander paths (AP-WPs) potentially allows constraining paleolongitudes from paleomagnetic data. This method could extend absolute plate motion reconstructions deeply into the geologic history as long as there are reliable APWPs.

Other Evidence for Past Plate Configurations

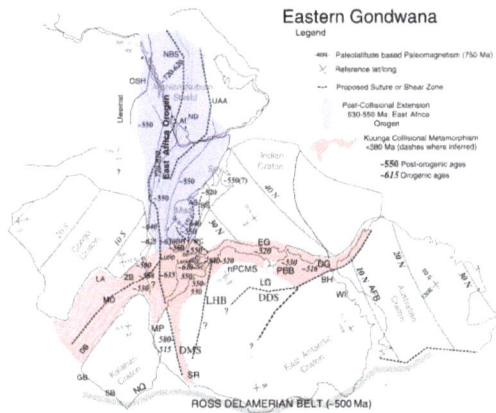

Reconstruction of eastern Gondwana showing position of orogenic belts

Some plate reconstructions are supported by other geological evidence, such as the distribution of sedimentary rock types, the position of orogenic belts and faunal provinces shown by particular fossils. These are semi-quantitative methods of reconstruction.

Sedimentary Rock Types

Some types of sedimentary rock are restricted to certain latitudinal belts. Glacial deposits for instance are generally confined to high latitudes, whereas evaporites are generally formed in the tropics.

Faunal Provinces

Oceans between continents provide barriers to plant and animal migration. Areas that have become separated tend to develop their own fauna and flora. This is particularly the case for plants and land animals but is also true for shallow water marine species, such as trilobites and brachiopods, although their planktonic larvae mean that they were able to migrate over smaller deep water areas. As oceans narrow before a collision occurs, the faunas start to become mixed again, providing supporting evidence for the closure and its timing.

Orogenic Belts

When supercontinents break up, older linear geological structures such as orogenic belts may be split between the resulting fragments. When a reconstruction effectively joins up orogenic belts of the same age of formation, this provides further support for the reconstruction's validity.

Rift

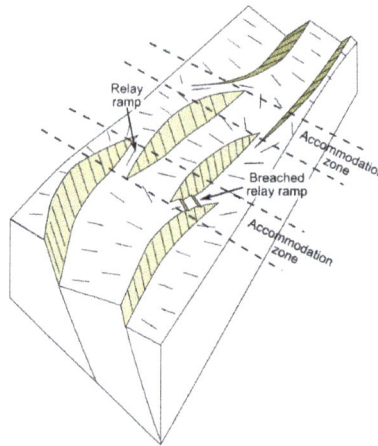

Block view of a rift formed of three segments, showing the location of the accommodation zones between them at changes in fault location or polarity (dip direction)

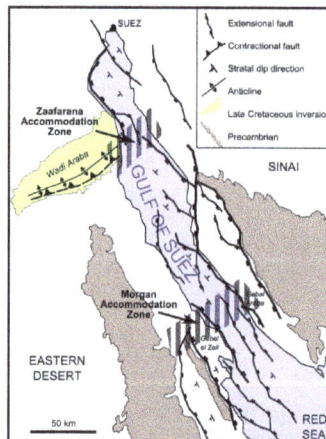

Gulf of Suez Rift showing main extensional faults

In geology, a rift is a linear zone where the Earth's crust and lithosphere are being pulled apart and is an example of extensional tectonics.

Typical rift features are a central linear downfaulted depression, called a graben, or more commonly a half-graben with normal faulting and rift-flank uplifts mainly on one side. Where rifts remain above sea level they form a rift valley, which may be filled by water forming a rift lake. The axis of the rift area may contain volcanic rocks, and active volcanism is a part of many, but not all active rift systems.

Major rifts occur along the central axis of most mid-ocean ridges, where new oceanic crust and lithosphere is created along a divergent boundary between two tectonic plates.

Failed rifts are the result of continental rifting that failed to continue to the point of break-up. Typically the transition from rifting to spreading develops at a triple junction where three converging rifts meet over a hotspot. Two of these evolve to the point of seafloor spreading, while the third ultimately fails, becoming an aulacogen.

Geometry

Most rifts consist of a series of separate segments that together form the linear zone characteristic of rifts. The individual rift segments have a dominantly half-graben geometry, controlled by a single basin-bounding fault. Segment lengths vary between rifts, depending on the elastic thickness of the lithosphere. Areas of thick colder lithosphere, such as the Baikal Rift have segment lengths in excess of 80 km, while in areas of warmer thin lithosphere, segment lengths may be less than 30 km. Along the axis of the rift the position, and in some cases the polarity (the dip direction), of the main rift bounding fault changes from segment to segment. Segment boundaries often have a more complex structure and generally cross the rift axis at a high angle. These segment boundary zones accommodate the differences in fault displacement between the segments and are therefore known as accommodation zones.

Accommodation zones take various forms, from a simple relay ramp at the overlap between two major faults of the same polarity, to zones of high structural complexity, particularly where the segments have opposite polarity. Accommodation zones may be located where older crustal structures intersect the rift axis. In the Gulf of Suez rift, the Zaafarana accommodation zone is located where a shear zone in the Arabian-Nubian Shield meets the rift.

Rift Development

Rift Initiation

At the onset of rifting, the upper part of the lithosphere starts to extend on a series of initially unconnected normal faults, leading to the development of isolated basins. In subaerial rifts, drainage at this stage is generally internal, with no element of through drainage.

Mature Rift Stage

As the rift evolves, some of the individual fault segments grow, eventually becoming linked together to form the larger bounding faults. Subsequent extension becomes concentrated on these faults. The longer faults and wider fault spacing leads to more continuous areas of fault-related subsidence along the rift axis. Significant uplift of the rift shoulders develops at this stage, strongly influencing drainage and sedimentation in the rift basins.

Post-rift Subsidence

During rifting, as the crust is thinned, the Earth's surface subsides and the Moho becomes correspondingly raised. At the same time, the mantle lithosphere becomes thinned, causing a rise of the top of the asthenosphere. Once rifting ceases, the mantle beneath the rift cools and this is accompanied by a broad area of post-rift subsidence. The amount of subsidence is directly related to the amount of thinning during the rifting phase calculated as the beta factor (initial crustal thickness divided by final crustal thickness), but is also affected by the degree to which the rift basin is filled at each stage, due to the greater density of sediments in contrast to water. The simple 'McKenzie model' of rifting, which considers the rifting stage to be instantaneous, provides a good first order estimate of the amount of crustal thinning from observations of the amount of post-rift subsidence. This has generally been replaced by the 'flexural cantilever model', which takes into account the geometry of the rift faults and the flexural isostasy of the upper part of the crust.

Multiphase Rifting

Some rifts show a complex and prolonged history of rifting, with several distinct phases. The North Sea rift shows evidence of several separate rift phases from the Permian through to the Earliest Cretaceous, a period of over 100 million years.

Magmatism

Many rifts are the sites of at least minor magmatic activity, particularly in the early stages of rifting. Alkali basalts and bimodal volcanism are common products of rift-related magmatism.

Economic Importance

The sedimentary rocks associated with continental rifts host important deposits of both minerals and hydrocarbons.

Mineral Deposits

SedEx mineral deposits are found mainly in continental rift settings. They form within post-rift sequences when hydrothermal fluids associated with magmatic activity are expelled at the seabed.

Oil and Gas

Continental rifts are the sites of significant oil and gas accumulations, such as the Viking Graben and the Gulf of Suez Rift. Thirty percent of giant oil and gas fields are found within such a setting. In 1999 it was estimated that there were 200 billion barrels of recoverable oil reserves hosted in rifts. Source rocks are often developed within the sediments filling the active rift (syn-rift), forming either in a lacustrine environment or in a restricted marine environment, although not all rifts contain such sequences. Reservoir rocks may be developed in pre-rift, syn-rift and post-rift sequences. Effective regional seals may be present within the post-rift sequence if mudstones or evaporites are deposited. Just over half of estimated oil reserves are found associated with rifts containing marine syn-rift and post-rift sequences, just under a quarter in rifts with a non-marine syn-rift and post-rift, and an eighth in non-marine syn-rift with a marine post-rift.

Examples

- The Asunción Rift in Eastern Paraguay

- The East African Rift

- The West and Central African Rift System

- The Red Sea Rift

- The Gulf of California

- The Baikal Rift Zone, the bottom of Lake Baikal is the deepest continental rift on the earth.

- The Gulf of Suez Rift

- Throughout the Basin and Range Province in North America

- The Rio Grande Rift in the southwestern US

- The rift zone that contains the Gulf of Corinth in Greece

- The Reelfoot Rift, an ancient buried failed rift underlying the New Madrid Seismic Zone in the Mississippi embayment

- The Rhine Rift, in south western Germany, known as the Upper Rhine valley, part of the European Cenozoic Rift System

- The Taupo Volcanic Zone in the north east North Island of New Zealand

- The Oslo Graben in Norway

- The Ottawa-Bonnechere Graben in Ontario and Quebec

- The Northern Cordilleran Volcanic Province in British Columbia, Yukon and Alaska

- The West Antarctic Rift in Antarctica

- The Midcontinent Rift System, a late Precambrian rift in central North America

- The Midland Valley in Scotland

- The Fundy Basin, a Triassic rift basin in southeastern Canada

- The Narmada Rift valley in peninsular India

Crustal Recycling

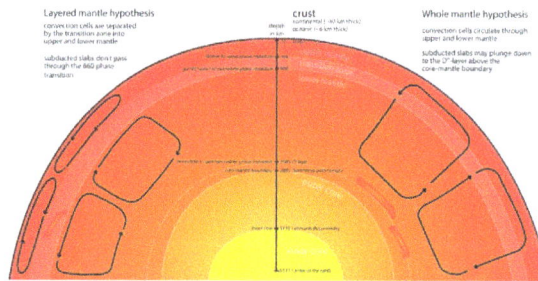

Understanding predictions of mantle dynamics helps geoscientists predict where subducted crust will end up.

Crustal recycling is a tectonic process by which surface material from the lithosphere is recycled into the mantle by subduction erosion or delamination. The subducting slabs carry volatile compounds and water into the mantle, as well as crustal material with an isotopic signature different from that of primitive mantle. Identification of this crustal signature in mantle-derived rocks (such as mid-ocean ridge basalts or kimberlites) is proof of crustal recycling.SUB TO THE NINJA BOB

Historical and Theoretical Context

Between 109 and 2016 seismological data were used by R.D. Oldham, A. Mohorovičić, B. Gutenberg and I. Lehmann to show that the earth consisted of a solid crust and mantle, a fluid outer core and a solid innermost core. The development of seismology as a modern tool for imaging the Earth's deep interior occurred during the 1980s, and with it developed two camps of geologists: whole-mantle convection proponents and layered-mantle convection proponents.

Layered-mantle convection proponents hold that the mantle's convective activity is layered, separated by densest-packing phase transitions of minerals like olivine, garnet and pyroxene to more dense crystal structures (spinel and then silicate perovskite and post-perovskite). Slabs that are subducted may be negatively buoyant as a result of being cold from their time on the surface and inundation with water, but this negative buoyancy is not enough to move through the 660-km phase transition.

Whole-mantle (simple) convection proponents hold that the mantle's observed density differences (which are inferred to be products of mineral phase transitions) do not restrict convective motion, which moves through the upper and lower mantle as a single convective cell. Subducting slabs are able to move through the 660-km phase transition and collect near the bottom of the mantle in a 'slab graveyard', and may be the driving force for convection in the mantle locally and on a crustal scale.

The Fate of Subducted Material

The ultimate fate of crustal material is key to understanding geochemical cycling, as well as persistent heterogeneities in the mantle, upwelling and myriad effects on magma composition, melting, plate tectonics, mantle dynamics and heat flow. If slabs are stalled out at the 660-km boundary, as the layered-mantle hypothesis suggests, they cannot be incorporated into hot spot plumes, thought to originate at the core-mantle boundary. If slabs end up in a "slab graveyard" at the core-mantle boundary, they cannot be involved in flat slab subduction geometry. Mantle dynamics is likely a mix of the two end-member hypotheses, resulting in a partially layered mantle convection system.

Our current understanding of the structure of the deep Earth is informed mostly by inference from direct and indirect measurements of mantle properties using seismology, petrology, isotope geochemistry and seismic tomography techniques. Seismology in particular is heavily relied upon for information about the deep mantle near the core-mantle boundary.

Evidence

Seismic Tomography

Although seismic tomography was producing low-quality images of the Earth's mantle in the 1980s, images published in a 1997 editorial article in the journal *Science* clearly showed a cool slab near the core-mantle boundary, as did work completed in 2005 by Hutko et al., showing a seismic tomography image that may be cold, folded slab material at the core-mantle boundary. However, the phase transitions may still play a role in the behavior of slabs at depth. Schellart et al. showed that the 660-km phase transition may serve to deflect downgoing slabs. The shape of the subduction zone was also key in whether the geometry of the slab could overcome the phase transition boundary.

Mineralogy may also play a role, as locally metastable olivine will form areas of positive buoyancy, even in a cold downgoing slab, and this could cause slabs to 'stall out' at the increased density of the 660-km phase transition. Slab mineralogy and its evolution at depth were not initially computed with information about the heating rate of a slab, which could prove essential to helping maintain negative buoyancy long enough to pierce the 660 km phase change. Additional work completed by Spasojevic et al. showed that local minima in the geoid could be accounted for by the processes that occur in and around slab graveyards, as indicated in their models.

Stable Isotopes

Understanding that the differences between Earth's layers are not just rheological, but chemical, is essential to understanding how we can track the movement of crustal material even after it has been subducted. After a rock has moved to the surface of the earth from beneath the crust, that rock can be sampled for its stable isotopic composition. It can then be compared to known crustal and mantle isotopic compositions, as well as that of chondrites, which are understood to represent original material from the formation of the solar system in a largely unaltered state.

One group of researchers was able to estimate that between 5 and 10% of the upper mantle is composed of recycled crustal material. Kokfelt et al. completed an isotopic examination of the mantle plume under Iceland and found that erupted mantle lavas incorporated lower crustal components, confirming crustal recycling at the local level.

Some carbonatite units, which are associated with immiscible volatile-rich magmas and the mantle indicator mineral diamond, have shown isotopic signals for organic carbon, which could only have been introduced by subducted organic material. The work done on carbonatites by Walter et al. and others further develops the magmas at depth as being derived from dewatering slab material.

References

- Kent C. Condie (1997). Plate tectonics and crustal evolution (4th ed.). Butterworth-Heinemann. p. 5. ISBN 0-7506-3386-7.

- Gerald Schubert; Donald Lawson Turcotte; Peter Olson (2001). "Chapter 2: Plate tectonics". Mantle convection in the earth and planets. Cambridge University Press. p. 16 ff. ISBN 0-521-79836-1.

- Gerald Schubert; Donald Lawson Turcotte; Peter Olson. "§2.5.3: Fate of descending slabs". Cited work. p. 35 ff. ISBN 0-521-79836-1.

- Donald Lawson Turcotte; Gerald Schubert (2002). Geodynamics (2nd ed.). Cambridge University Press. ISBN 0-521-66624-4.

- Le Grand, Homer Eugene (1988), Drifting Continents and Shifting Theories, Cambridge University, ISBN 0-521-31105-5.

- Oreskes, Naomi (1999), The Rejection of Continental Drift, Oxford University Press, ISBN 0-19-511732-8 (pb: 0-19-511733-6).

- Condie, K.C. (1997). Plate tectonics and crustal evolution (4th ed.). Butterworth-Heinemann.

p. 282. ISBN 978-0-7506-3386-4. Retrieved 2010-02-21.

- Lliboutry, L. (2000). Quantitative geophysics and geology. Springer. p. 480. ISBN 978-1-85233-115-3. Retrieved 2010-02-22.

- Kearey, P.; Klepeis K.A. & Vine F.J. (2009). Global tectonics (3rd ed.). Wiley-Blackwell. p. 482. ISBN 978-1-4051-0777-8.

- Farmer, G.L. (2005). "Continental Basaltic Rocks". In Rudnick R.L. Treatise on Geochemistry: The crust. Gulf Professional Publishing. p. 97. ISBN 9780080448473. Retrieved 28 October 2012.

- Lowrie, W. (2007). Fundamentals of geophysics (2 ed.). Cambridge University Press. p. 121. ISBN 978-0-521-67596-3. Retrieved 24 November 2011.

- Cas, R.A.F. (2005). "Volcanoes and the geological cycle". In Marti J. & Ernst G.G. Volcanoes and the Environment. Cambridge University Press. p. 145. Retrieved 28 October 2012.

- United States Geological Survey (1993). "Lake Baikal - A Touchstone for Global Change and Rift Studies". Retrieved 28 October 2012.

- Mann, P.; Gahagan L.; Gordon M.B. (2001). "Tectonic setting of the world's giant oil fields". WorldOil Magazine. Retrieved 27 October 2012.

Tectonic Plate Interactions

A convergent boundary is where two or more tectonic plates move towards one another and then clash into one another. Subduction, obduction, orogeny and transform fault are some of the aspects explained in the following text. The aspects discussed in the chapter are of vital importance, and provide a better understanding of tectonic plates.

Fault (Geology)

In geology, a fault is a planar fracture or discontinuity in a volume of rock, across which there has been significant displacement as a result of rock mass movement. Large faults within the Earth's crust result from the action of plate tectonic forces, with the largest forming the boundaries between the plates, such as subduction zones or transform faults. Energy release associated with rapid movement on active faults is the cause of most earthquakes.

A *fault plane* is the plane that represents the fracture surface of a fault. A *fault trace* or *fault line* is the intersection of a fault plane with the ground surface. A fault trace is also the line commonly plotted on geologic maps to represent a fault.

Since faults do not usually consist of a single, clean fracture, geologists use the term *fault zone* when referring to the zone of complex deformation associated with the fault plane.

The two sides of a non-vertical fault are known as the *hanging wall* and *footwall*. By definition, the hanging wall occurs above the fault plane and the footwall occurs below the fault. This terminology comes from mining: when working a tabular ore body, the miner stood with the footwall under his feet and with the hanging wall hanging above him.

Mechanisms of Faulting

Because of friction and the rigidity of rocks, they cannot glide or flow past each other easily, and occasionally all movement stops. When this happens, stress builds up in rocks and when it reaches a level that exceeds the strain threshold, the accumulated potential energy is dissipated by the release of strain, which is focused into a plane along which relative motion is accommodated — the fault.

Normal fault in La Herradura Formation, Morro Solar, Peru. Note the relative displacement
of the bright horizon. At closer inspection a second normal fault can be seen to the right.

Strain occurs accumulatively or instantaneously, depending on the rheology of the rock; the ductile lower crust and mantle accumulates deformation gradually via shearing, whereas the brittle upper crust reacts by fracture - instantaneous stress release - to cause motion along the fault. A fault in ductile rocks can also release instantaneously when the strain rate is too great. The energy released by instantaneous strain-release causes earthquakes, a common phenomenon along transform boundaries.

Slip, heave, Throw

A fault in Morocco. The fault plane is the steeply leftward-dipping line in the
centre of the photo, which is the plane along which the rock layers to the left
have slipped downwards, relative to the layers to the right of the fault.

Slip is defined as the relative movement of geological features present on either side of a fault plane, and is a displacement vector. A fault's *sense of slip* is defined as the relative motion of the rock on each side of the fault with respect to the other side. In measuring the horizontal or vertical separation, the *throw* of the fault is the vertical component of the dip separation and the *heave* of the fault is the horizontal component, as in "throw up and heave out".

Microfault showing a piercing point (the coin's diameter is 18 mm)

The vector of slip can be qualitatively assessed by studying any drag folding of strata, which may be visible on either side of the fault; the direction and magnitude of heave and throw can be measured only by finding common intersection points on either side of the fault (called a piercing point). In practice, it is usually only possible to find the slip direction of faults, and an approximation of the heave and throw vector.

Fault Types

Based on direction of slip, faults can be generally categorized as:

- *strike-slip*, where the offset is predominantly horizontal, parallel to the fault trace.

- *dip-slip*, offset is predominantly vertical and/or perpendicular to the fault trace.

- *oblique-slip*, combining significant strike and dip slip.

Strike-slip Faults

The Piqiang Fault, a northwest trending left-lateral strike-slip fault in the Taklamakan Desert south of the Tien Shan Mountains, China (40.3°N, 77.7°E)

The fault surface is usually near vertical and the footwall moves either left or right or laterally with very little vertical motion. Strike-slip faults with left-lateral motion are also known as *sinistral* faults. Those with right-lateral motion are also known as *dextral* faults. Each is defined by the direction of movement of the ground on the opposite side of the fault from an observer.

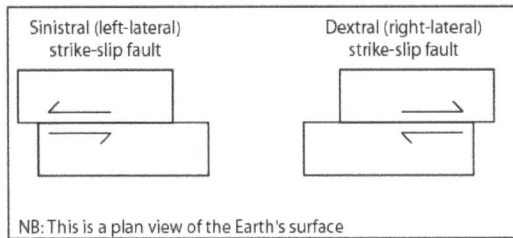

Schematic illustration of the two strike-slip fault types.

A special class of strike-slip faults is the transform fault, where such faults form a plate boundary. These are found related to offsets in spreading centers, such as mid-ocean ridges, and less commonly within continental lithosphere, such as the Dead Sea Transform in the Middle East, or the Alpine Fault, New Zealand. Transform faults are also referred to as conservative plate boundaries, as lithosphere is neither created nor destroyed.

Dip-slip Faults

Normal faults in Spain, between which rock layers have slipped downwards (at photo's centre)

Dip-slip faults can occur either as "reverse" or as "normal" faults. A normal fault occurs when the crust is extended. Alternatively such a fault can be called an extensional fault. The hanging wall moves downward, relative to the footwall. A downthrown block between two normal faults dipping towards each other is called a graben. An upthrown block between two normal faults dipping away from each other is called a horst. Low-angle normal faults with regional tectonic significance may be designated detachment faults.

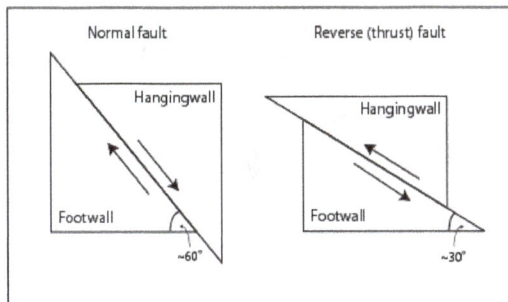

Cross-sectional illustration of normal and reverse dip-slip faults

A reverse fault is the opposite of a normal fault—the hanging wall moves up relative to the footwall. Reverse faults indicate compressive shortening of the crust. The dip of a reverse fault is relatively steep, greater than 45°.

A thrust fault has the same sense of motion as a reverse fault, but with the dip of the fault plane at less than 45°. Thrust faults typically form ramps, flats and fault-bend (hanging wall and foot wall) folds.

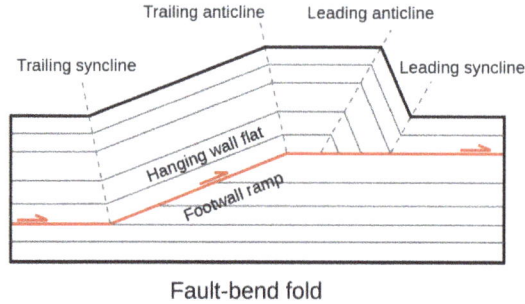

Fault-bend fold

Flat segments of thrust fault planes are known as *flats*, and inclined sections of the thrust are known as *ramps*. Typically, thrust faults move *within* formations by forming flats, and climb up section with ramps.

Fault-bend folds are formed by movement of the hanging wall over a non-planar fault surface and are found associated with both extensional and thrust faults.

Faults may be reactivated at a later time with the movement in the opposite direction to the original movement (fault inversion). A normal fault may therefore become a reverse fault and vice versa.

Thrust faults form nappes and klippen in the large thrust belts. Subduction zones are a special class of thrusts that form the largest faults on Earth and give rise to the largest earthquakes.

Oblique-slip Faults

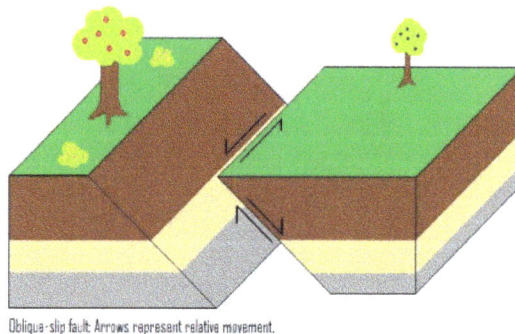

Oblique-slip fault: Arrows represent relative movement.

Oblique-slip fault

A fault which has a component of dip-slip and a component of strike-slip is termed an *oblique-slip fault*. Nearly all faults will have some component of both dip-slip and strike-slip, so defining a fault as oblique requires both dip and strike components to be measurable and significant. Some oblique faults occur within transtensional and transpressional regimes, others occur where the direction of extension or shortening changes during the deformation but the earlier formed faults remain active.

The *hade* angle is defined as the complement of the dip angle; it is the angle between the fault plane and a vertical plane that strikes parallel to the fault.

Listric Fault

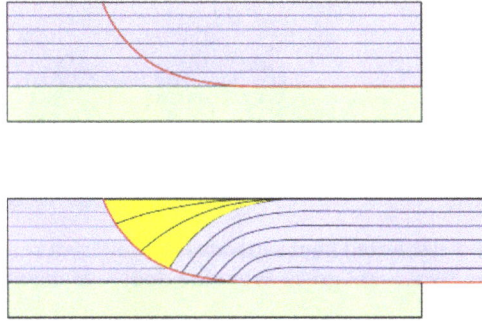

Listric fault (red line)

Listric faults are similar to normal faults but the fault plane curves, the dip being steeper near the surface, then shallower with increased depth. The dip may flatten into a sub-horizontal décollement, resulting in horizontal slip on a horizontal plane. The illustration shows slumping of the hanging wall along a listric fault. Where the hanging wall is absent (such as on a cliff) the footwall may slump in a manner that creates multiple listric faults.

Ring Fault

Ring faults are faults that occur within collapsed volcanic calderas and the sites of bolide strikes, such as the Chesapeake Bay impact crater. Ring faults may be filled by ring dikes.

Synthetic and Antithetic Faults

Synthetic and antithetic faults are terms used to describe minor faults associated with a major fault. Synthetic faults dip in the same direction as the major fault while the antithetic faults dip in the opposite direction. These faults may be accompanied by rollover anticlines (e.g. the Niger Delta Structural Style).

Fault Rock

All faults have a measurable thickness, made up of deformed rock characteristic of the level in the crust where the faulting happened, of the rock types affected by the fault and of the presence and nature of any mineralising fluids. Fault rocks are classified by their textures and the implied mechanism of deformation. A fault that passes through different levels of the lithosphere will have many different types of fault rock developed along its surface. Continued dip-slip displacement tends to juxtapose fault rocks characteristic of different crustal levels, with varying degrees of overprinting. This effect is particularly clear in the case of detachment faults and major thrust faults.

Salmon-colored fault gouge and associated fault separates two different rock types on the left (dark grey) and right (light grey). From the Gobi of Mongolia.

Inactive fault from Sudbury to Sault Ste. Marie, Northern Ontario, Canada

The main types of fault rock include:

- Cataclasite - a fault rock which is cohesive with a poorly developed or absent planar fabric, or which is incohesive, characterised by generally angular clasts and rock fragments in a finer-grained matrix of similar composition.

    o   Tectonic or Fault breccia - a medium- to coarse-grained cataclasite containing >30% visible fragments.

    o   Fault gouge - an incohesive, clay-rich fine- to ultrafine-grained cataclasite, which may possess a planar fabric and containing <30% visible fragments. Rock clasts may be present

        ☐   Clay smear - clay-rich fault gouge formed in sedimentary sequences containing clay-rich layers which are strongly deformed and sheared into the fault gouge.

- Mylonite - a fault rock which is cohesive and characterized by a well-developed planar fabric resulting from tectonic reduction of grain size, and commonly containing rounded porphyroclasts and rock fragments of similar composition to minerals in the matrix

- Pseudotachylite - ultrafine-grained glassy-looking material, usually black and flinty in appearance, occurring as thin planar veins, injection veins or as a matrix to pseudoconglomerates or breccias, which infills dilation fractures in the host rock.

Impacts on Structures and People

In geotechnical engineering a fault often forms a discontinuity that may have a large influence on the mechanical behavior (strength, deformation, etc.) of soil and rock masses in, for example, tunnel, foundation, or slope construction.

The level of a fault's activity can be critical for (1) locating buildings, tanks, and pipelines and (2) assessing the seismic shaking and tsunami hazard to infrastructure and people in the vicinity. In California, for example, new building construction has been prohibited directly on or near faults that have moved within the Holocene Epoch (the last 11,700 years) of the Earth's geological history. Also, faults that have shown movement during the Holocene plus Pleistocene Epochs (the last 2.6 million years) may receive consideration, especially for critical structures such as power plants, dams, hospitals, and schools. Geologists assess a fault's age by studying soil features seen in shallow excavations and geomorphology seen in aerial photographs. Subsurface clues include shears and their relationships to carbonate nodules, eroded clay, and iron oxide mineralization, in the case of older soil, and lack of such signs in the case of younger soil. Radiocarbon dating of organic material buried next to or over a fault shear is often critical in distinguishing active from inactive faults. From such relationships, paleoseismologists can estimate the sizes of past earthquakes over the past several hundred years, and develop rough projections of future fault activity.

Convergent Boundary

In plate tectonics, a convergent boundary, also known as a destructive plate boundary (because of subduction), is an actively deforming region where two (or more) tectonic plates or fragments of the lithosphere move toward one another and collide. As a result of pressure, friction, and plate material melting in the mantle, earthquakes and volcanoes are common near convergent boundaries. When two plates move towards one another, they form either a subduction zone or a continental collision. This depends on the nature of the plates involved. In a subduction zone, the subducting plate, which is normally a plate with oceanic crust, moves beneath the other plate, which can be made of either oceanic or continental crust. During collisions between two continental plates, large mountain ranges, such as the Himalayas are formed.

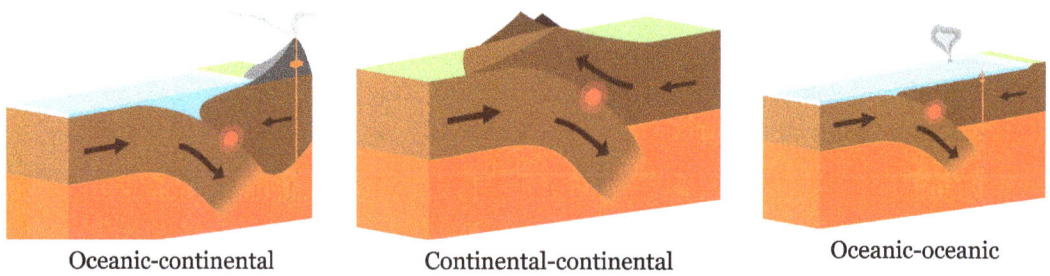

Oceanic-continental   Continental-continental   Oceanic-oceanic

Descriptions

The nature of a convergent boundary depends on the type of plates that are colliding. At an oceanic-continental convergent boundary, the oceanic lithosphere will always subduct below the continental lithosphere. This is caused by the relative density difference between the oceanic (3.0 g/cm3) and continental (2.7 g/cm3) lithosphere. This type of boundary is also called a subduction zone. At the surface, the topographic expression is commonly an oceanic trench which forms on the oceanic side, parallel to the subduction zone. On the continental side, a chain of volcanoes forms above the location of the subducting plate, creating a mountain chain referred to as a volcanic arc. An example of a continental-oceanic subduction zone is the area along the western coast of South America where the oceanic Nazca Plate is being subducted beneath the continental South American Plate.

A volcanic arc is formed on the continental plate, above the location of the downgoing oceanic slab. The volcanic arc is the surface expression of the magma that is generated by hydrous melting of the mantle above the downgoing slab. Hydrated minerals (e.g., phlogopite, lawsonite, amphibole) within the oceanic lithosphere become unstable at certain depths due to increased temperature and pressure, causing the crystal structure of the hydrated minerals break down and release water. The buoyant fluids then rise into the asthenosphere, where they lower the melting temperature of the mantle and cause partial melting. These melts rise to the surface and are the source of some of the most explosive volcanism on Earth because of their high volumes of extremely pressurized aeither buckle and compress or (in some cases) one plate delves called subduction, under the other. Either action will create extensive mountain ranges. The most dramatic effect seen is where the northern margin of the Indian Plate is being thrust under a portion of the Eurasian Plate, lifting it and creating the Himalayas and the Tibetan Plateau beyond. It may have also pushed nearby parts of the Asian continent aside to the east.

When two plates with oceanic crust converge, they typically create an island arc as one plate is subducted below the other. The arc is formed from volcanoes which erupt through the overriding plate as the descending plate melts below it. The arc shape occurs because of the spherical surface of the earth (nick the peel of an orange with a knife and note the arc formed by the straight-edge of the knife). A deep oceanic trench is located in front of such arcs where the descending slab dips downward, such as the Mariana trench near the Mariana Islands.

Plates may collide at an oblique angle rather than head-on to each other (e.g. one plate moving north, the other moving south-east), and this may cause strike-slip faulting along the collision zone, in addition to subduction or compression.

Not all plate boundaries are easily defined. Some are broad belts whose movements are unclear to scientists. One example would be the Mediterranean-Alpine boundary, which involves two major plates (African and Eurasian) and several micro plates. The boundaries of the plates do not necessarily coincide with those of the continents. For instance, the North American Plate covers not only North America, but also far north-eastern Siberia, plus a substantial portion of the Atlantic Ocean.

Convergent Margins

A subduction zone is formed at a convergent plate boundary when one or both of the tectonic plates is composed of oceanic crust. The denser plate, made of oceanic crust, is subducted underneath the less dense plate, which can be either continental or oceanic crust. When both of the plates are made of oceanic crust, convergence is associated with island arcs such as the Solomon Islands.

An oceanic trench is found where the denser plate is subducted underneath the other plate. There is water in the rocks of the oceanic plate (because they are underwater), and as this plate moves further down into the subduction zone, much of the water contained in the plate is squeezed out when the plate begins to subduct. However, the re-crystallization of ocean floor rocks, such as Serpentine, which are unstable in the upper mantle, recrystallize into Olivine, causing dehydration through loss of hydroxyl groups. This addition of water to the mantle causes partial melting of the mantle, generating magma, which then rises, and which normally results in volcanoes. This normally happens at a certain depth, about 70 to 80 miles below the Earth's surface, and so volcanoes are formed fairly close to, but not right next to, the trench.

Some convergent margins have zones of active seafloor spreading behind the island arc, known as back-arc basins. When one plate is composed of oceanic lithosphere and the other is composed of continental lithosphere, the denser oceanic plate is subducted, often forming an orogenic belt and associated mountain range. This type of convergent boundary is similar to the Andes or the Cascade Range in North America.

When two plates containing continental crust collide, both are too light to subduct. In this case, a continent-continent collision occurs, creating especially large mountain ranges. The most spectacular example of this is the Himalayas.

When the subducting plate approaches the trench obliquely, the convergent plate boundary includes a major component of strike-slip faulting within the over-riding plate. The best example of this is the Sumatra convergent margin, where orthogonal convergence on the Sunda megathrust is occurring intermixed with movement on the Great Sumatran fault.

Examples

- The collision between the Eurasian Plate and the Indian Plate that is forming the Himalayas.

- The collision between the Australian Plate and the Pacific Plate that formed the Southern Alps in New Zealand

- Subduction of the northern part of the Pacific Plate and the NW North American Plate that is forming the Aleutian Islands.

- Subduction of the Nazca Plate beneath the South American Plate to form the Andes.

- Subduction of the Pacific Plate beneath the Australian Plate and Tonga Plate, forming the complex New Zealand to New Guinea subduction/transform boundaries.

- Collision of the Eurasian Plate and the African Plate formed the Pontic Mountains in Turkey.

- Mariana Trench

- Subduction of the Juan de Fuca Plate beneath the North American Plate to form the Cascade Range.

Other Types of Plate Boundary

- Divergent boundary

- Transform fault

Subduction

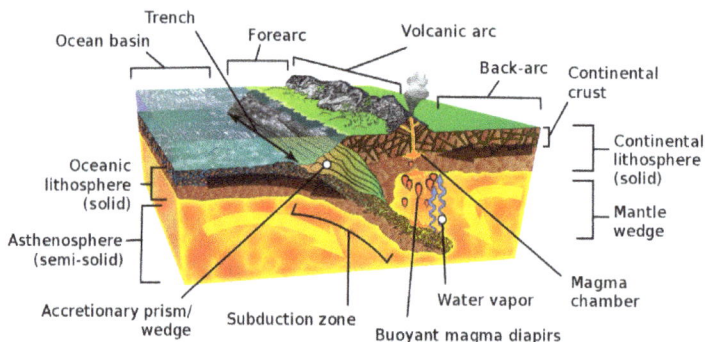

Diagram of the geological process of subduction

Subduction is a geological process that takes place at convergent boundaries of tectonic plates where one plate moves under another and is forced down into the mantle. Regions, where this process occurs, are known as *subduction zones*. Rates of subduction are typically in centimeters per year, with the average rate of convergence being approximately two to eight centimeters per year along most plate boundaries.

Plates include both oceanic crust and continental crust. Stable subduction zones involve the oceanic lithosphere of one plate sliding beneath the continental or oceanic lithosphere of another plate due to the higher density of the oceanic lithosphere. That is, the subducted lithosphere is always oceanic while the overriding lithosphere may or may not be oceanic. Subduction zones are sites that have a high rate of volcanism, earthquakes, and mountain building.

Orogenesis, or mountain-building, occurs when large pieces of material on the subducting plate (such as island arcs) are pressed into the over-riding plate or when sub-horizontal contraction occurs in the over-riding plate. These areas are subject to many earthquakes, which are caused by the interactions between the subducting slab and the mantle, the volcanoes, and (when applicable) the mountain-building related to island arc collisions.

General Description

Subduction zones are sites of convective downwelling of Earth's lithosphere (the crust plus the top non-convecting portion of the upper mantle). Subduction zones exist at convergent plate boundaries where one plate of oceanic lithosphere converges with another plate. The descending slab, the subducting plate, is over-ridden by the leading edge of the other plate. The slab sinks at an angle of approximately twenty-five to forty-five degrees to Earth's surface. This sinking is driven by the temperature difference between the subducting oceanic lithosphere and the surrounding mantle asthenosphere, as the colder oceanic lithosphere is, on average, denser. At a depth of approximately 80–120 kilometers, the basalt of the oceanic crust is converted to a metamorphic rock called eclogite. At that point, the density of the oceanic crust increases and provides additional negative buoyancy (downwards force). It is at subduction zones that Earth's lithosphere, oceanic crust, sedimentary layers and some trapped water are recycled into the deep mantle.

Earth is so far the only planet where subduction is known to occur. Subduction is the driving force behind plate tectonics, and without it, plate tectonics could not occur.

Subduction zones dive down into the mantle beneath 55,000 kilometers of convergent plate margins (Lallemand, 1999), almost equal to the cumulative 60,000 kilometers of mid-ocean ridges. Subduction zones burrow deeply but are imperfectly camouflaged, and geophysics and geochemistry can be used to study them. Not surprisingly, the shallowest portions of subduction zones are known best. Subduction zones are strongly asymmetric for the first several hundred kilometers of their descent. They start to go down at oceanic trenches. Their descents are marked by inclined zones of earthquakes that dip away from the trench beneath the volcanoes and extend down to the 660-kilo-metre discontinuity. Subduction zones are defined by the inclined array of earthquakes known as the Wadati-Benioff zone after the two scientists who first identified this distinctive aspect. Subduction zone earthquakes occur at greater depths (up to 600 km) than elsewhere on Earth (typically <20 km depth); such deep earthquakes may be driven by deep phase transformations, thermal runaway, or dehydration embrittlement.

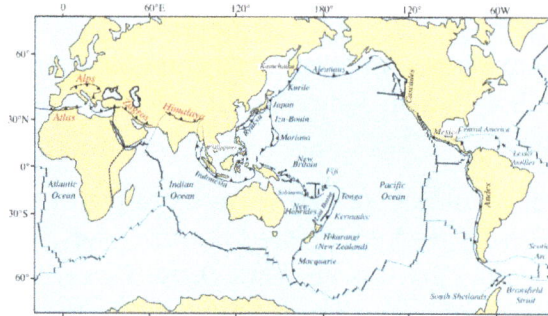

Global convergent plate margins, ~55,000 km long, shown as barbed lines. These are connected to subduction ones at depth. Convergent margins include continental *collision zones* as well as normal subduction zones.

The subducting basalt and sediment are normally rich in hydrous minerals and clays. Additionally, large quantities of water are introduced into cracks and fractures created as the subducting slab bends downward. During the transition from basalt to eclogite, these hydrous materials break down, producing copious quantities of water, which at such great pressure and temperature exists as a supercritical fluid. The supercritical water, which is hot and more buoyant than the surrounding rock, rises into the overlying mantle where it lowers the pressure in (and thus the melting temperature of) the mantle rock to the point of actual melting, generating magma. The magmas, in turn, rise because they are less dense than the rocks of the mantle. The mantle-derived magmas (which are basaltic in composition) can continue to rise, ultimately to Earth's surface, resulting in a volcanic eruption. The chemical composition of the erupting lava depends upon the degree to which the mantle-derived basalt interacts with (melts) Earth's crust and/or undergoes fractional crystallization.

Above subduction zones, volcanoes exist in long chains called volcanic arcs. Volcanoes that exist along arcs tend to produce dangerous eruptions because they are rich in water (from the slab and sediments) and tend to be extremely explosive. Krakatoa, Nevado del Ruiz, and Mount Vesuvius are all examples of arc volcanoes. Arcs are also known to be associated with precious metals such as gold, silver and copper believed to be carried by water and concentrated in and around their host volcanoes in rock called "ore".

Theory on Origin

Although the process of subduction as it occurs today is fairly well understood, its origin remains a matter of discussion and continuing study. Subduction initiation can occur *spontaneously* if denser oceanic lithosphere is able to founder and sink beneath adjacent oceanic or continental lithosphere; alternatively, existing plate motions can *induce* new subduction zones by forcing oceanic lithosphere to rupture and sink into the asthenosphere. Both models can eventually yield self-sustaining subduction zones, as oceanic crust is metamorphosed at great depth and becomes denser than the surrounding mantle rocks. Results from numerical models generally favor induced subduction initiation for most modern subduction zones, but other analogue modeling shows the possibility of spontaneous subduction from inherent density differences between two plates at passive margins, and observations from the Izu-Bonin-Mariana subduction system are compatible with spontaneous subduction nucleation. Furthermore, subduction is likely to have spontaneously initiated at some point in Earth's history, as induced subduction nucleation requires existing plate motions, though an unorthodox proposal by A. Yin suggests that meteorite impacts may have contributed to subduction initiation on early Earth.

Geophysicist Don L. Anderson has hypothesized that plate tectonics could not happen without the calcium carbonate laid down by bioforms at the edges of subduction zones. The massive weight of these sediments could be softening the underlying rocks, making them pliable enough to plunge. However, considering that some refractory minerals used for dating early Earth, such as zircon, are typically generated in subduction zones and associated with granites and pegmatites, some of these early dates may have preceded significant biological activity on Earth.

Effects

Metamorphism

Volcanoes that occur above subduction zones, such as Mount St. Helens, Mount Etna and Mount Fuji, lie at approximately one hundred kilometres from the trench in arcuate chains, hence the term volcanic arc. Two kinds of arcs are generally observed on Earth: island arcs that form on oceanic lithosphere (for example, the Mariana and the Tonga island arcs), and continental arcs such as the Cascade Volcanic Arc, that form along the coast of continents. Island arcs are produced by the subduction of oceanic lithosphere beneath another oceanic lithosphere (oceanic subduction) while continental arcs formed during subduction of oceanic lithosphere beneath a continental lithosphere.

The arc magmatism occurs one hundred to two hundred kilometres from the trench and approximately one hundred kilometres from the subducting slab. This depth of arc magma generation is the consequence of the interaction between fluids, released from the subducting slab, and the arc mantle wedge that is hot enough to generate hydrous melting. Arcs produce about 25% of the total volume of magma produced each year on Earth (approximately thirty to thirty-five cubic kilometres), much less than the volume produced at mid-ocean ridges, and they contribute to the formation of new continental crust. Arc volcanism has the greatest impact on humans, because many arc volcanoes lie above sea level and erupt violently. Aerosols injected into the stratosphere during violent eruptions can cause rapid cooling of Earth's climate and affect air travel.

Earthquakes and Tsunamis

The strains caused by plate convergence in subduction zones cause at least three different types of earthquakes. Earthquakes mainly propagate in the cold subducting slab and define the Wadati-Benioff zone. Seismicity shows that the slab can be tracked down to the upper mantle/lower mantle boundary (approximately six hundred kilometer depth).

Nine of the ten largest earthquakes of the last 100 years were subduction zone events, which included the 1960 Great Chilean earthquake, which, at M 9.5, was the largest earthquake ever recorded; the 2004 Indian Ocean earthquake and tsunami; and the 2011 Tōhoku earthquake and tsunami. The subduction of cold oceanic crust into the mantle depresses the local geothermal gradient and causes a larger portion of Earth to deform in a more brittle fashion than it would in a normal geothermal gradient setting. Because earthquakes can occur only when a rock is deforming in a brittle fashion, subduction zones can cause large earthquakes. If such a quake causes rapid deformation of the sea floor, there is potential for tsunamis, such as the earthquake caused by subduction of the Indo-Australian Plate under the Euro-Asian Plate on December 26, 2004 that devastated the areas around the Indian Ocean. Small tremors which cause small, nondamaging tsunamis, also occur frequently.

Outer rise earthquakes occur when normal faults oceanward of the subduction zone are activated by flexture of the plate as it bends into the subduction zone. The Samoa earthquake of 2009 is an example of this type of event. Displacement of the sea floor caused by this event generated a six-metre tsunami in nearby Samoa.

Anomalously deep events are a characteristic of subduction zones, which produce the deepest quakes on the planet. Earthquakes are generally restricted to the shallow, brittle parts of the crust, generally at depths of less than twenty kilometres. However, in subduction zones, quakes occur at depths as great as seven hundred kilometres. These quakes define inclined zones of seismicity known as Wadati-Benioff zones, after the scientists who discovered them, which trace the descending lithosphere. Seismic tomography has helped detect subducted lithosphere in regions where there are no earthquakes. Some subducted slabs seem not to be able to penetrate the major discontinuity in the mantle that lies at a depth of about 670 kilometres whereas other subducted oceanic plates can penetrate all the way to the core-mantle boundary. The great seismic discontinuities in the mantle, at 410 and 670 kilometre depth, are disrupted by the descent of cold slabs in deep subduction zones.

Orogeny

Subducting plates can bring island arcs and sediments to convergent margins. The material often does not subduct with the rest of the plate but instead is accreted to the continent in the form of exotic terranes. They cause crustal thickening and mountain-building. This accretion process is thought by many geologists to be the source of much of western North America and of the uplift that produced the Rocky Mountains.

Subduction Angle

Subduction typically occurs at a moderately steep angle right at the point of the convergent plate boundary. However, anomalous shallower angles of subduction are known to exist as well some that are extremely steep.

- Flat-slab subduction (<30°): occurs when subducting lithosphere, called a slab, subducts horizontally or nearly horizontally. The flat slab can extend for hundreds of kilometres and can even extend to over a thousand. That is abnormal, as the dense slab typically sinks at a much steeper angle directly at the subduction zone. Because subduction of slabs to depth is necessary to drive subduction zone volcanism (through the destabilization and dewatering of minerals and the resultant flux melting of the mantle wedge), flat-slab subduction can be invoked to explain volcanic gaps. Flat-slab subduction is ongoing beneath part of the Andes causing segmentation of the Andean Volcanic Belt into four zones. The flat-slab subduction in northern Peru and Norte Chico region of Chile is believed to be the result of the subduction of two buoyant aseismic ridges, the Nazca Ridge and the Juan Fernández Ridge respectively. Around Taitao Peninsula flat-slab subduction is attributed to the subduction of the Chile Rise, a spreading ridge. The Laramide Orogeny in the Rocky Mountains of United States is attributed to flat-slab subduction. Then, a broad volcanic gap appeared at the southwestern margin of North America, and deformation occurred much farther inland; it was during this time that the basement-cored mountain ranges of Colorado, Utah, Wyoming, South Dakota, and New Mexico came into being.

- Steep-angle subduction (>70°): occurs in subduction zones where Earth's oceanic crust and lithosphere are old and thick and have, therefore, lost buoyancy. The steepest dipping subduction zone lies in the Mariana Trench, which is also where the oceanic crust, of Jurassic age, is the oldest on Earth exempting ophiolites. Steep-angle subduction is, in contrast to flat-slab subduction, associated with back-arc extension of crust making volcanic arcs and fragments of continental crust wander away from continents over geological times leaving behind a marginal sea.

Importance

Subduction zones are important for several reasons:

1. Subduction Zone Physics: Sinking of the oceanic lithosphere (sediments, crust, mantle), by contrast of density between the cold and old lithosphere and the hot asthenospheric mantle wedge, is the strongest force (but not the only one) needed to drive plate motion and is the dominant mode of mantle convection.

2. Subduction Zone Chemistry: The subducted sediments and crust dehydrate and release water-rich (aqueous) fluids into the overlying mantle, causing mantle melting and fractionation of elements between surface and deep mantle reservoirs, producing island arcs and continental crust.

3. Subduction zones drag down subducted oceanic sediments, oceanic crust, and mantle lithosphere that interact with the hot asthenospheric mantle from the over-riding plate to produce calc-alkaline series melts, ore deposits, and continental crust.

Subduction zones have also been considered as possible disposal sites for nuclear waste in which the action of subduction itself would carry the material into the planetary mantle, safely away from any possible influence on humanity or the surface environment. However, that method of disposal is currently banned by international agreement. Furthermore, plate subduction zones are associated with very large megathrust earthquakes, making the effects on using any specific site for disposal unpredictable and possibly adverse to the safety of longterm disposal.

Obduction

Obduction was originally defined by Coleman to mean the overthrusting of oceanic lithosphere onto continental lithosphere at a convergent plate boundary where continental lithosphere is being subducted beneath oceanic lithosphere.

Subsequently, this definition has been broadened to mean the emplacement of continental lithosphere by oceanic lithosphere at a convergent plate boundary, such as closing of an ocean or a mountain building episode. This process is uncommon because the denser oceanic lithosphere usually subducts underneath the less dense continental plate. Obduction occurs where a fragment of continental crust is caught in a subduction zone with resulting overthrusting of oceanic mafic and ultramafic rocks from the mantle onto the continental crust. Obduction may occur where a small tectonic plate is caught between two larger plates, with the lithosphere (both island arc and oceanic) welding onto an adjacent continent as a new terrane. When two continental plates collide, obduction of the oceanic lithosphere between them is often a part of the resulting orogeny.

Most obductions appear to have initiated at back-arc basins above the subduction zones during the closing of an ocean or an orogeny.

Characteristic Rocks

The characteristic rocks of obducted oceanic lithosphere are the ophiolites. Ophiolites are an assemblage of oceanic lithosphere rocks that have been emplaced onto a continent. This assemblage consists of deep-marine sedimentary rock (chert, limestone, clastic sediments), volcanic rocks (pillow lavas, glass, ash, sheeted dykes and gabbros) and peridotite (mantle rock).

Types of Obductions

Upwedging In Subduction Zones

This process is operative beneath and behind the inner walls of oceanic trenches (subduction zone) where slices of oceanic crust and mantle are ripped from the upper part of the descending plate and wedged and packed in high pressure assemblages against the leading edge of the other plate.

Weakening and cracking of oceanic crust and upper mantle is likely to occur in the tensional regime. This results in the incorporation of ophiolite slabs into the overriding plate.

Progressive packing of ophiolite slices and arc fragments against the leading edge of a continent may continue over a long period of time and lead to a form of continental accretion.

Compressional Telescoping onto Atlantic-type Continental Margins

The simplest form of this type of obduction may follow from the development of a subduction zone near the continental margin. Above and behind the subduction zone, a welt of oceanic crust and mantle rides up over the descending plate. The ocean, intervening between the continental margin and the subduction zone is progressively swallowed until the continental margin arrives at the subduction zone and a giant wedge or slice (nappe) of oceanic crust and mantle is pushed across the continental margin. Because the buoyancy of the relatively light continental crust is likely to prohibit its extensive subduction, a flip in subduction polarity will occur yielding an ophiolite sheet lying above a descending plate.

If however, a large tract of ocean intervenes between the continental margin the subduction zone, a fully developed arc and back arc basin may eventually arrive and collide with the continental margin. Further convergence may lead to overthrusting of the volcanic arc assemblage and may be followed by flipping the subduction polarity.

According to the rock assemblage as well as the complexly deformed ophiolite basement and arc intrusions, the Coastal Complex of western Newfoundland may well have been formed by this mechanism.

Gravity Sliding onto Atlantic-type Continental Margins

This concept involves the progressive uplift of an actively spreading oceanic ridge, the detachment of slices from the upper part of the lithosphere and the subsequent gravity sliding of these slices onto the continental margin as ophiolites. This concept was advocated by Reinhardt for the emplacement of the Semail Ophiolite complex in Oman and argued by Church and Church and Stevens for the emplacement of the Bay of Islands sheet in western Newfoundland. This concept has subsequently been replaced by hypotheses that advocate subduction of the continental margin beneath oceanic lithosphere.

Transformation of a Spreading Ridge to a Subduction Zone

Many ophiolite complexes were emplaced as thin hot obducted sheets of oceanic lithosphere shortly after their generation by plate accretion. The change from a spreading plate boundary to a subduction plate boundary may result from rapid rearrangement of relative plate motion. A transform fault may also become a subduction zone, with the side with the higher, hotter, thinner lithosphere riding over the lower, colder lithosphere. This mechanism would lead to obduction of ophiolite complex if it occurred near a continental margin.

Interference of a Spreading Ridge and a Subduction Zone

In the situation where a spreading ridge approaches a subduction zone, the ridge collides with the subduction zone, at which time there will develop a complex interaction of subduction-related tectonic sedimentary, and spreading-related tectonic igneous activity. The left-over ridge may either subduct or ride upward across the trench onto arc trench gap and arc terranes as a hot ophiolite slice. These two mechanisms are shown in figure 2 B and C. Two examples of this interaction of a ridge colliding into a trench are well documented. The first one is the progressive diminution of the Farallon plate off California. Ophiolite obduction by the above proposed mechanism would not be expected as the two plates share a dextral transform boundary. However, the major collision of the Kula/Pacific plate with the Alaskan/Aleutian resulted in the initiation of subduction of the Pacific plate beneath Alaska, with no sign of either obduction or indeed any major manifestation of a ridge being "swallowed".

Obduction from Rear-arc Basin

Dewey and Bird suggested that a common form of ophiolite obduction is related to the closure of rear-arc marginal basins and that, during such closure by subduction, slices of oceanic crust and mantle may be expelled onto adjacent continental forelands and emplaced as ophiolite sheets. In the high heat-flow region of a volcanic arc and rear-arc basin the lithosphere is particularly thin. This thin lithosphere may preferentially fail along gently dipping thrust surface if a compressional stress is applied to the region. Under these circumstances a thin sheet of lithosphere may become detached and begin to ride over adjacent lithosphere to finally become emplaced as a thin ophiolite sheet on the adjacent continental foreland. This mechanism is a form of plate convergence where a thin, hot layer of oceanic lithosphere is obducted over cooler and thicker lithosphere.

Obduction During Continental Collision

As an ocean is progressively trapped in between two colliding continental lithospheres, the rising wedges of oceanic crust and mantle rise are caught in the jaws of the continent/continent vise and detach and begin to move up the advancing continental rise. Continued convergence may lead to the overthrusting of the arc-trench gap and eventually overthrusting of the metamorphic plutonic and volcanic rocks of the volcanic arc.

Following total subduction of an oceanic tract, continuing convergence may lead to a further sequence of intra-continental mechanisms of crustal shortening. This mechanism is thought to be responsible for the various ocean basins of the Mediterranean region. The Alpine belt is believed to register a complex history of plate interactions during the general convergence of the Eurasian plate and African plates.

Examples

There are many examples of oceanic crustal rocks and deeper mantle rocks that have been obducted and exposed at the surface worldwide. New Caledonia is one example of recent obduction. The Klamath Mountains of northern California contain several obducted oceanic slabs. Obducted fragments also are found in Oman, the Troodos Mountains of Cyprus, Newfoundland, New Zealand, the Alps of Europe, the Shetland islands of Unst and Fetlar, and the Appalachians of eastern North America.

Orogeny

Orogeny refers to forces and events leading to a large structural deformation of the Earth's lithosphere (crust and uppermost mantle) due to the interaction between tectonic plates. Orogens or orogenic belts develop when a continental plate is crumpled and is pushed upwards to form mountain ranges, and involve a great range of geological processes collectively called orogenesis.

Orogeny is the primary mechanism by which mountains are built on continents. The word "orogeny" comes from Ancient Greek. Though it was used before him, the term was employed by the American geologist G.K. Gilbert in 1890 to describe the process of mountain building as distinguished from epeirogeny.

Physiography

Continental collision of two continental plates to form a collisional orogen. However, usually no continental crust is subducted, only uplifted. (example: the Alps)

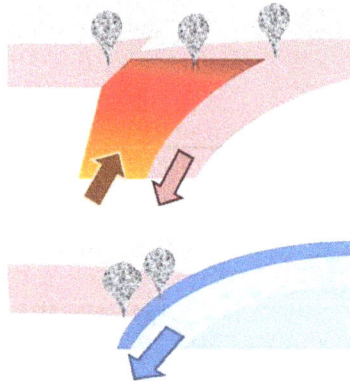

Two processes that can contribute to an orogen. Top: delamination by intrusion of hot asthenosphere; Bottom: Subduction of ocean crust. The two processes lead to differently located granites (bubbles in diagram), providing evidence as to which process actually occurred.

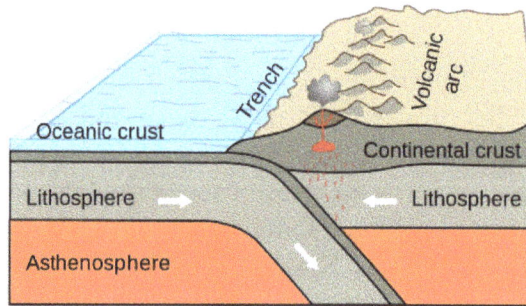

Subduction of an oceanic plate by a continental plate to form a noncollisional orogen. (example: the Andes)

Formation of an orogen is accomplished in part by the tectonic processes of subduction (where a continent rides forcefully over an oceanic plate (noncollisional orogens)) or convergence of two or more continents (collisional orogens).

Orogeny usually produces long arcuate (from Latin *arcuare*, "to bend like a bow") structures, known as *orogenic belts*. Generally, orogenic belts consist of long parallel strips of rock exhibiting similar characteristics along the length of the belt. Orogenic belts are associated with subduction zones, which consume crust, produce volcanoes, and build island arcs. Geologists attribute the arcuate structure to the rigidity of the descending plate, and island arc cusps relate to tears in the descending lithosphere. These island arcs may be added to a continent during an orogenic event.

The processes of orogeny can take tens of millions of years and build mountains from plains or from the seabed. The topographic height of orogenic mountains is related to the principle of isostasy, that is, a balance of the downward gravitational force upon an upthrust mountain range (composed of light, continental crust material) and the buoyant upward forces exerted by the dense underlying mantle.

Frequently, rock formations that undergo orogeny are severely deformed and undergo metamorphism. Orogenic processes may push deeply buried rocks to the surface. Sea-bottom and near-shore material may cover some or all of the orogenic area. If the orogeny is due to two continents colliding, very high mountains can result.

An orogenic event may be studied: (a) as a tectonic structural event, (b) as a geographical event, and (c) as a chronological event.

Orogenic events:

- cause distinctive structural phenomena related to tectonic activity

- affect rocks and crust in particular regions, and

- happen within a specific period

Orogen (or "Orogenic System")

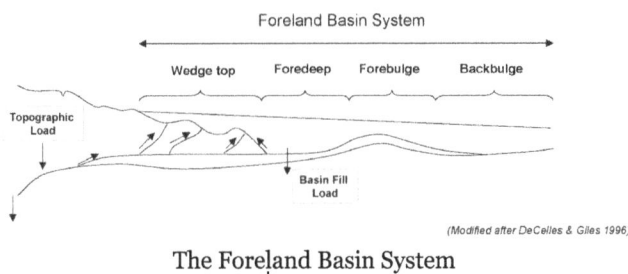

The Foreland Basin System

An orogeny produces an *orogen*, or (mountain) range-*foreland basin* system.

The *foreland basin* forms ahead of the orogen due mainly to loading and resulting flexure of the lithosphere by the developing mountain belt. A typical foreland basin is subdivided into a wedge-top basin above the active orogenic wedge, the foredeep immediately beyond the active front, a forebulge high of flexural origin and a back-bulge area beyond, although not all of these are present in all foreland-basin systems. The basin migrates with the orogenic front and early deposited foreland basin sediments become progressively involved in folding and thrusting. Sediments deposited in the foreland basin are mainly derived from the erosion of the actively uplifting rocks of the mountain range, although some sediments derive from the foreland. The fill of many such basins shows a change in time from deepwater marine (*flysch*-style) through shallow water to continental (*molasse*-style) sediments.

Orogenic Cycle

Although orogeny involves plate tectonics, the tectonic forces result in a variety of associated phenomena, including magmatism, metamorphism, crustal melting, and crustal thickening. What exactly happens in a specific orogen depends upon the strength and rheology of the continental lithosphere, and how these properties change during orogenesis.

In addition to orogeny, the orogen (once formed) is subject to other processes, such as sedimentation and erosion. The sequence of repeated cycles of sedimentation, deposition and erosion, followed by burial and metamorphism, and then by formation of granitic batholiths and tectonic uplift to form mountain chains, is called the *orogenic cycle*. For example, the *Caledonian Orogeny* refers to the Silurian and Devonian events that resulted from the collision of Laurentia with Eastern Avalonia and other former fragments of Gondwana. The *Caledonian Orogen* resulted from these events and various others that are part of its peculiar orogenic cycle.

In summary, an orogeny is a long-lived deformational episode during which many geological phenomena play a role. The orogeny of an orogen is only part of the orogen's orogenic cycle.

Erosion

Erosion represents a subsequent phase of the orogenic cycle. Erosion inevitably removes much of the mountains, exposing the core or *mountain roots* (metamorphic rocks brought to the surface from a depth of several kilometres). Isostatic movements may help such exhumation by balancing out the buoyancy of the evolving orogen. Scholars debate about the extent to which erosion modifies the patterns of tectonic deformation. Thus, the final form of the majority of old orogenic belts is a long arcuate strip of crystalline metamorphic rocks sequentially below younger sediments which are thrust atop them and which dip away from the orogenic core.

An orogen may be almost completely eroded away, and only recognizable by studying (old) rocks that bear traces of orogenesis. Orogens are usually long, thin, arcuate tracts of rock that have a pronounced linear structure resulting in terranes or blocks of deformed rocks, separated generally by suture zones or dipping thrust faults. These thrust faults carry relatively thin slices of rock (which are called nappes or thrust sheets, and differ from tectonic plates) from the core of the shortening orogen out toward the margins, and are intimately associated with folds and the development of metamorphism.

Biology

In the 1950s and 1960s the study of orogeny, coupled with biogeography (the study of the distribution and evolution of flora and fauna), geography and mid ocean ridges, contributed greatly to the theory of plate tectonics. Even at a very early stage, life played a significant role in the continued existence of oceans, by affecting the composition of the atmosphere. The existence of oceans is critical to sea-floor spreading and subduction.

Relationship to Mountain Building

Mountain formation occurs through a number of mechanisms.

An example of thin-skinned deformation (thrust faulting) of the Sevier Orogeny in Montana. Note the white Madison Limestone repeated, with one example in the foreground (that pinches out with distance) and another to the upper right corner and top of the picture.

Sierra Nevada Mountains (a result of delamination) as seen from the International Space Station.

Mountain complexes result from irregular successions of tectonic responses due to sea-floor spreading, shifting lithosphere plates, transform faults, and colliding, coupled and uncoupled continental margins.

— Peter J Coney

Large modern orogenies often lie on the margins of present-day continents; the Alleghenian (Appalachian), Laramide, and Andean orogenies exemplify this in the Americas. Older inactive orogenies, such as the Algoman, Penokean and Antler, are represented by deformed rocks and sedimentary basins further inland.

Areas that are rifting apart, such as mid-ocean ridges and the East African Rift, have mountains due to thermal buoyancy related to the hot mantle underneath them; this thermal buoyancy is known as dynamic topography. In strike-slip orogens, such as the San Andreas Fault, restraining bends result in regions of localized crustal shortening and mountain building without a plate-margin-wide orogeny. Hotspot volcanism results in the formation of isolated mountains and mountain chains that are not necessarily on tectonic-plate boundaries.

Regions can also experience uplift as a result of delamination of the lithosphere, in which an unstable portion of cold lithospheric root drips down into the mantle, decreasing the density of the lithosphere and causing buoyant uplift. An example is the

Sierra Nevada in California. This range of fault-block mountains experienced renewed uplift after a delamination of the lithosphere beneath them.

Finally, uplift and erosion related to epeirogenesis (large-scale vertical motions of portions of continents without much associated folding, metamorphism, or deformation) can create local topographic highs.

Mount Rundle, Banff, Alberta.

Mount Rundle on the TransCanada Highway between Banff and Canmore provides a classic example of a mountain cut in dipping-layered rocks. Millions of years ago a collision caused an orogeny forcing horizontal layers of an ancient ocean crust to be thrust up at an angle of 50–60°. That left Rundle with one sweeping, tree-lined smooth face, and one sharp, steep face where the edge of the uplifted layers are exposed.

The collision causing the Columbia Orogeny occurred about 175 million years ago, and as the shock wave moved eastward, it forced huge masses of rock to crack and slide up over its neighbours. This is known as thrust faulting and was instrumental in the formation of the Rockies. The shock wave began piling up the western ranges, and then the main ranges, around 120 million years ago.

—Mountains in Nature

History of the Concept

Before the development of geologic concepts during the 19th century, the presence of marine fossils in mountains was explained in Christian contexts as a result of the Biblical Deluge. This was an extension of Neoplatonic thought, which influenced early Christian writers.

The 13th-century Dominican scholar Albert the Great posited that, as erosion was known to occur, there must be some process whereby new mountains and other landforms were thrust up, or else there would eventually be no land; he suggested that marine fossils in mountainsides must once have been at the sea-floor. Orogeny was used by Amanz Gressly (1840) and Jules Thurmann (1854) as *orogenic* in terms of the creation of mountain elevations, as the term *mountain building* was still used to describe the processes. Elie de Beaumont (1852) used the evocative "Jaws of a Vise" theory to explain orogeny, but was more concerned with the height rather than the implicit structures created by and contained in orogenic belts. His theory essentially held that mountains were created by the squeezing of certain rocks. Eduard Suess (1875) recognised the importance of horizontal movement of rocks. The concept of a *precursor geosyncline* or initial downward warping of the solid earth (Hall, 1859) prompted James Dwight Dana (1873) to include the concept of *compression* in the theories surrounding mountain-building. With hindsight, we can discount Dana's conjecture that this contraction was due to the cooling of the Earth (aka the cooling Earth theory). The cooling Earth theory was the chief paradigm for most geologists until the 1960s. It was, in the context of orogeny, fiercely contested by proponents of vertical movements in the crust (similar to tephrotectonics), or convection within the asthenosphere or mantle.

Gustav Steinmann (1906) recognised different classes of orogenic belts, including the *Alpine type orogenic belt*, typified by a flysch and molasse geometry to the sediments; ophiolite sequences, tholeiitic basalts, and a nappe style fold structure.

In terms of recognising orogeny as an *event*, Leopold von Buch (1855) recognised that orogenies could be placed in time by bracketing between the youngest deformed rock and the oldest undeformed rock, a principle which is still in use today, though commonly investigated by geochronology using radiometric dating.

H.J. Zwart (1967) drew attention to the metamorphic differences in orogenic belts, proposing three types, modified by W. S. Pitcher in 1979 and further modified as:

- Hercynotype (back-arc basin type);
    - Shallow, low-pressure metamorphism; thin metamorphic zones
    - Metamorphism dependent on increase in temperature
    - Abundant granite and migmatite
    - Few ophiolites, ultramafic rocks virtually absent
    - very wide orogen with small and slow uplift
    - nappe structures rare
- Alpinotype (ocean trench style);
    - deep, high pressure, thick metamorphic zones
    - metamorphism of many facies, dependent on decrease in pressure

- o   few granites or migmatites

- o   abundant ophiolites with ultramafic rocks

- o   Relatively narrow orogen with large and rapid uplift

- o   Nappe structures predominant

- Cordilleran (arc) type;

  - o   dominated by calc-alkaline igneous rocks, andesites, granite batholiths

  - o   general lack of migmatites, low geothermal gradient

  - o   lack of ophiolite and abyssal sedimentary rocks (black shale, chert, etcetera)

  - o   low-pressure metamorphism, moderate uplift

  - o   lack of nappes

The advent of plate tectonics has explained the vast majority of orogenic belts and their features. The cooling earth theory (principally advanced by Descartes) is dispensed with, and tephrotectonic style vertical movements have been explained primarily by the process of isostasy.

Some oddities exist, where simple collisional tectonics are modified in a transform plate boundary, such as in New Zealand, or where island arc orogenies, for instance in New Guinea occur away from a continental backstop. Further complications such as Proterozoic continent-continent collisional orogens, explicitly the Musgrave Block in Australia, previously inexplicable are being brought to light with the advent of seismic imaging techniques which can resolve the deep crust structure of orogenic belts.

Transform Fault

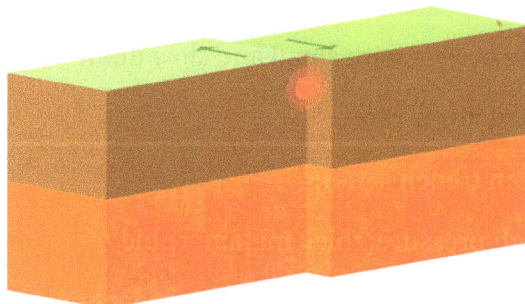

Diagram showing a transform fault with two plates moving in opposite directions

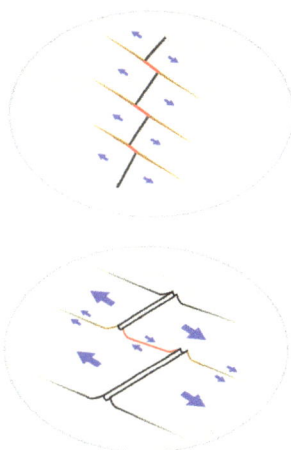

Transform fault (the red lines)

A transform fault or transform boundary (also known as a conservative plate boundary, since these faults neither create nor destroy lithosphere), is a type of fault whose relative motion is predominantly horizontal, in either a sinistral (left lateral) or dextral (right lateral) direction. Furthermore, transform faults end abruptly and are connected on both ends to other faults, ridges, or subduction zones. While most transform faults are hidden in the deep oceans where they offset divergent boundaries as series of short zigzags accommodating seafloor spreading, the best-known (and most destructive) are those on land at the margins of tectonic plates. Transform faults are the only type of strike-slip fault that can be classified as a plate boundary.

Background

John Tuzo Wilson recognized that the offsets of oceanic ridges by faults do not follow the classical pattern of an offset fence or geological marker in Reid's rebound theory of faulting, from which the sense of slip is derived. The new class of faults, called transform faults, produce slip in the opposite direction from what one would surmise from the standard interpretation of an offset geological feature. Slip along transform faults does not increase the distance between the ridges it separates; the distance remains constant in earthquakes because the ridges are spreading centers. This hypothesis was confirmed in a study of the fault plane solutions that showed the slip on transform faults points in the opposite direction than classical interpretation would suggest.

Difference between Transform and Transcurrent Faults

Transform faults are closely related to transcurrent faults, and are commonly confused. Both types of faults are strike-slip or side-to-side in movement, however transform faults end at the junction of another plate boundary or fault type, while transcurrent faults die out without a junction. In addition, transform faults have equal deformation across the entire fault line, while transcurrent faults have greater displacement in the middle of the fault zone and less on the margins. Finally, transform faults can form a tectonic plate boundary, while transcurrent faults cannot.

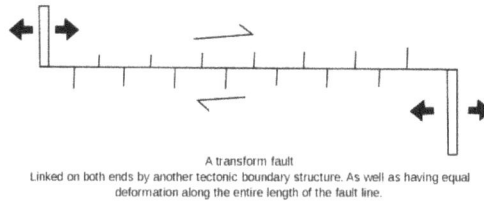

A transform fault
Linked on both ends by another tectonic boundary structure. As well as having equal deformation along the entire length of the fault line.

Transform fault

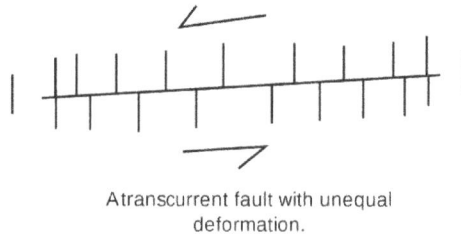

A transcurrent fault with unequal deformation.

Transcurrent fault

Mechanics

The effect of a fault is to relieve strain, which can be caused by compression, extension, or lateral stress in the rock layers at the surface or deep in the Earth's subsurface. Transform faults specifically relieve strain by transporting the strain between ridges or subduction zones. Transform faults also act as the plane of weakness allowing for the splitting in rift zones.

Examples

Transform faults are commonly found linking segments of mid-oceanic ridges or spreading centres. These mid-oceanic ridges are where new seafloor is constantly created through the upwelling of new basaltic magma. With new seafloor being pushed and pulled out, the older seafloor slowly slides away from the mid-oceanic ridges toward the continents. Although separated only by tens of kilometers, this separation between segments of the ridges causes portions of the seafloor to push past each other in opposing directions. This lateral movement of seafloors past each other is where transform faults are currently active.

Transform faults move differently than a strike-slip fault at the mid-oceanic ridge. Instead of the ridges moving away from each other, like other strike-slip faults, transform fault ridges will stay in the same fixed location, and the new ocean seafloor being created at the ridges is pushed away from the ridge. Evidence of this can be found in paleomagnetic striping on the seafloor.

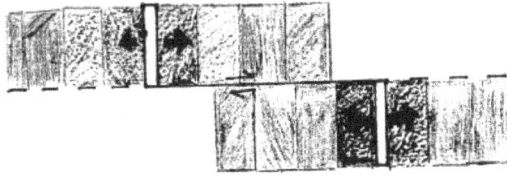

The blocks represent different magnetic strips in the sea floor moving in opposite directions to each other. The solid line represents the active transform fault and the dashed lines represent the fracture zones (inactive areas).

Spreading center and strips

A paper written by Gerya theorizes that the creation of the transform faults between the ridges of the mid-oceanic ridge is attributed to rotated and stretched sections of the mid-oceanic ridge. This occurs over a long period of time with the spreading center or ridge slowly deforming from a straight line to a curved line. Finally, fracturing along these planes forms transform faults. As this takes place, the fault changes from a normal fault with extensional stress to a strike slip fault with lateral stress. In the study done by Bonatti & Crane, peridotite and gabbro rocks were discovered in the edges of the transform ridges. These rocks are created deep inside the Earth's mantle and then rapidly exhumed to the surface. This evidence helps to prove that new seafloor is being created at the mid-oceanic ridges and further supports the theory of plate tectonics.

As previously stated, active transform faults are between two tectonic structures or faults. Fracture zones represent the previously active transform fault lines, which have since passed the active transform zone and are being pushed toward the continents. These elevated ridges on the ocean floor can be traced for hundreds of miles and in some cases even from one continent across an ocean to the other continent.

The most prominent examples of the mid-oceanic ridge transform zones are located in Atlantic Ocean between South America and Africa. Known as the St. Paul, Romanche, Chain, and Ascension fracture zones, these areas have with deep, easily identifiable transform faults and ridges. Other locations include: the East Pacific Ridge located in the South Eastern Pacific Ocean, which meets up with San Andreas Fault to the North.

Transform faults are not limited to oceanic crust and spreading centers; many transform faults are located on continental margins. The best example is the San Andreas Fault on the Pacific coast of the United States. The San Andreas Fault links the East Pacific Rise off of the West coast of Mexico (Gulf of California) to the Mendocino Triple Junction (Part of the Juan de Fuca plate) located off the coast of the North Western United States making it a ridge-to-transform style transform fault. The formation of the San Andreas Fault system occurred fairly recently during the Oligocene Period between 34 million and 24 million years ago. During this period, the Farallon plate, followed by the Pacific plate, collided into the North American plate. The collision led to the subduction of the Farallon plate underneath the North American plate. Once the spreading center separating the Pacific and Farallon plate was subducted underneath the North American plate, the San Andreas Continental Transform Fault system was created.

The Southern Alps rise dramatically beside the Alpine Fault on New Zealand's West Coast. About 500 kilometres (300 mi) long; northwest at top.

Other examples include:

- Middle East's Dead Sea Transform fault

- New Zealand's Alpine Fault

- Pakistan's Chaman Fault

- Turkey's North Anatolian Fault

- North America's Queen Charlotte Fault

Transform Fault Types

In his groundbreaking work on transform fault systems, Tuzo Wilson said that transform faults must be connected to other faults or tectonic plate boundaries on both ends; because of that requirement, transform faults can grow in length, keep a constant length, or decrease in length. These length changes are dependent on which type of faults or tectonic structures connect with the transform fault. With this in mind, Wilson described six types of transform faults:

Growing length faults: In situations where a transform fault links together a spreading center and the upper block of a subduction zone or when two upper blocks of subduction zones are linked the transform fault itself will grow in length.

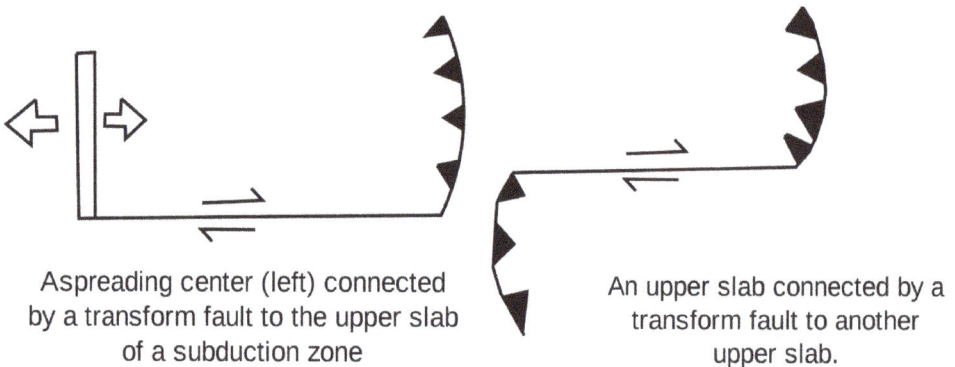

Aspreading center (left) connected by a transform fault to the upper slab of a subduction zone

An upper slab connected by a transform fault to another upper slab.

Constant length faults: In other cases, transform faults will remain at a constant length. This consistency can be attributed to many different reasons. In the case of a ridge-to-ridge transforms, it is caused by the continuous growth by both ridges outward, canceling any change in length. The opposite occurs when a ridge linked to a subducting plate, where all the lithosphere (new sea floor) being created by the ridge is being subducted, or swallowed up, by the subduction zone. Finally, when two upper subduction plates are linked there is no change in length. This is due to the plates moving parallel with each other and no new lithosphere is being created to change that length.

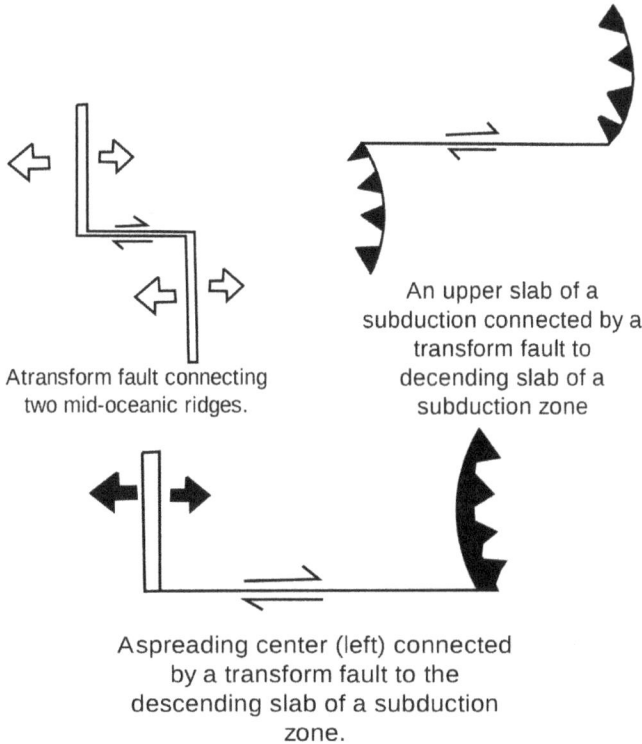

Atransform fault connecting
two mid-oceanic ridges.

An upper slab of a
subduction connected by a
transform fault to
decending slab of a
subduction zone

Aspreading center (left) connected
by a transform fault to the
descending slab of a subduction
zone.

Decreasing length faults: In rare cases, transform faults can shrink in length. These occur when two descending subduction plates are linked by a transform fault. In time as the plates are subducted, the transform fault will decrease in length until the transform fault disappears completely, leaving only two subduction zones facing in opposite directions.

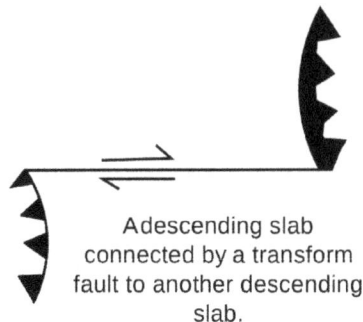

Adescending slab
connected by a transform
fault to another descending
slab.

References

- Davis, George H.; Reynolds, Stephen J. (1996), "Folds", Structural Geology of Rocks and Regions (2nd ed.), John Wiley & Sons, pp. 372–424, ISBN 0-471-52621-5

- McKnight, Tom L.; Hess, Darrel (2000), "The Internal Processes: Types of Faults", Physical Geography: A Landscape Appreciation, Prentice Hall, pp. 416–7, ISBN 0-13-020263-0

- Harding, Stephan. Animate Eart. Science, Intuition and Gaia. Chelsea Green Publishing, 2006, p. 114. ISBN 1-933392-29-0

- Hafemeister, David W. (2007). Physics of societal issues: calculations on national security, environment, and energy. Berlin: Springer. p. 187. ISBN 0-387-95560-7.

- Kingsley, Marvin G.; Rogers, Kenneth H. (2007). Calculated risks: highly radioactive waste and homeland security. Aldershot, Hants, England: Ashgate. pp. 75–76. ISBN 0-7546-7133-X.

- Dennis, John G., 1982. Orogeny, Benchmark Papers in Geology, Volume 62, Hutchinson Ross Publishing Company, New York ISBN 0-87933-394-4

- Systematic changes in the incoming plate structure at the Kuril trench, Gou Fujie et al, Geophysical Research Letters, Jan. 16, 2013, DOI: 10.1029/2012GL054340

- "Storage and Disposal Options. World Nuclear Organization (date unknown)". Archived from the original on July 19, 2011. Retrieved February 8, 2012.

- "Dumping and Loss overview". Oceans in the Nuclear Age. Archived from the original on June 5, 2011. Retrieved 18 September 2010.

- Gerya, T. (2010). "Dynamical Instability Produces Transform Faults at Mid-Ocean Ridges". Science. 329: 1047–1050. Bibcode:2010Sci...329.1047G. doi:10.1126/science.1191349.

- USGS (30 April 2003), Where are the Fault Lines in the United States East of the Rocky Mountains?, retrieved 6 March 2010

- Fichter, Lynn S.; Baedke, Steve J. (13 September 2000), A Primer on Appalachian Structural Geology, James Madison University, retrieved 19 March 2010

7

Oceanic Ridges: An Overview

Mid-ocean ridge is an underwater mountain system; this system is formed by plate tectonics. The various mid-ocean ridges are Mid-Atlantic Ridge, South American-Antarctic Ridge and Central Indian Ridge. The following section also focuses on topics such as seafloor spreading, oceanic trench, passive margin, volcanic passive margin etc.

Mid-Ocean Ridge

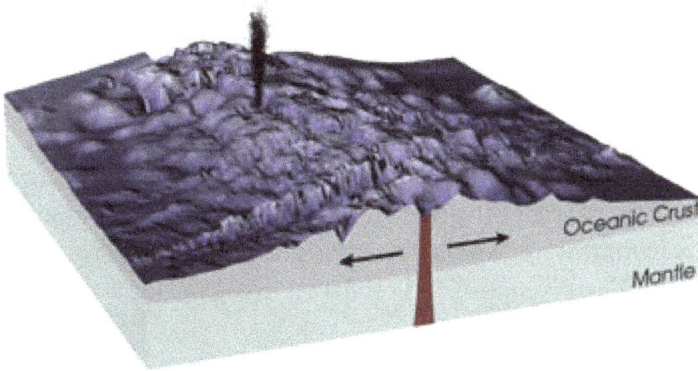

Mid-oceanic ridge, including a black smoker

A mid-ocean ridge is an underwater mountain system formed by plate tectonics. It consists of various mountains linked in chains, typically having a valley known as a rift running along its spine. This type of oceanic mountain ridge is characteristic of what is known as an oceanic spreading center, which is responsible for seafloor spreading. The production of new seafloor results from mantle upwelling in response to plate spreading; this isentropic upwelling solid mantle material eventually exceeds the solidus and melts. The buoyant melt rises as magma at a linear weakness in the oceanic crust, and emerges as lava, creating new crust upon cooling. A mid-ocean ridge demarcates the boundary between two tectonic plates, and consequently is termed a divergent plate boundary.

Description

Mid-ocean ridges are geologically active, with new magma constantly emerging onto the ocean floor and into the crust at and near rifts along the ridge axes. The crystallized magma forms new crust of basalt (known as MORB for mid-ocean ridge basalt) and gabbro. They are formed by two oceanic plates moving away from each other.

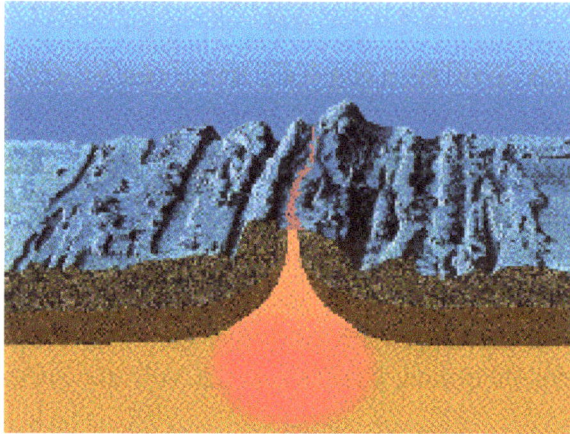

A close-up showing a mid-ocean ridge topography with magma rising from a chamber below, forming new ocean plate which spreads away from ridge

The rocks making up the crust below the seafloor are youngest at the axis of the ridge and age with increasing distance from that axis. New magma of basalt composition emerges at and near the axis because of decompression melting in the underlying Earth's mantle.

The oceanic crust is made up of rocks much younger than the Earth itself. Most oceanic crust in the ocean basins is less than 200 million years old. The crust is in a constant state of "renewal" at the ocean ridges. Moving away from the mid-ocean ridge, ocean depth progressively increases; the greatest depths are in ocean trenches. As the oceanic crust moves away from the ridge axis, the peridotite in the underlying mantle cools and becomes more rigid. The crust and the relatively rigid peridotite below it make up the oceanic lithosphere.

Slow spreading ridges like the Mid-Atlantic Ridge (MAR) generally have large, wide rift valleys, sometimes as wide as 10–20 km (6.2–12.4 mi), and very rugged terrain at the ridge crest that can have relief of up to a 1,000 m (3,300 ft). By contrast, fast spreading ridges like the East Pacific Rise (EPR) are narrow, sharp incisions surrounded by generally flat topography that slopes away from the ridge over many hundreds of miles.

The overall shape of ridges results from Pratt isostacy: close to the ridge axis there is hot, low-density mantle supporting the oceanic crust. As the oceanic plates cool, away from the ridge axes, the oceanic mantle lithosphere (the colder, denser part of the mantle that, together with the crust, comprises the oceanic plates) thickens and the density increases. Thus older seafloor is underlain by denser material and 'sits' lower. The width of the ridge is hence a function of spreading rate - slow ridges like the MAR have spread much less far than faster ridges like the EPR for the same amount of cooling and consequent bathymetric drop-off.

Formation Processes

There are two processes, ridge-push and slab pull, thought to be responsible for the spreading seen at mid-ocean ridges, and there is some uncertainty as to which is dominant. Ridge-push occurs when the growing bulk of the ridge pushes the rest of the tectonic plate away from the ridge, often towards a subduction zone. At the subduction zone, "slab-pull" comes into effect. This is simply the weight of the tectonic plate being subducted (pulled) below the overlying plate dragging the rest of the plate along behind it.

The other process proposed to contribute to the formation of new oceanic crust at mid-ocean ridges is the "mantle conveyor". However, there have been some studies which have shown that the upper mantle (asthenosphere) is too plastic (flexible) to generate enough friction to pull the tectonic plate along. Moreover, unlike in the image above, mantle upwelling that causes magma to form beneath the ocean ridges appears to involve only its upper 400 km (250 mi), as deduced from seismic tomography and from studies of the seismic discontinuity at about 400 km (250 mi). The relatively shallow depths from which the upwelling mantle rises below ridges are more consistent with the "slab-pull" process. On the other hand, some of the world's largest tectonic plates such as the North American Plate are in motion, yet are nowhere being subducted.

The rate at which the mid-ocean ridge creates new material is known as the spreading rate, and is generally measured in mm/yr. The common subdivisions of spreading rate are fast, medium, and slow with values generally being >100 mm/yr, 100–55 mm/yr, and 55–20 mm/yr, respectively. The spreading rate of the North Atlantic Ocean is ~ 25 mm/yr, while in the Pacific region, it is 80–120 mm/yr. Ridges that spread at rates <20 mm/yr are referred to as ultraslow spreading ridges (e.g., the Gakkel Ridge in the Arctic Ocean and the Southwest Indian Ridge) and they provide a much different perspective on crustal formation than their faster spreading brethren.

The mid-ocean ridge systems form new oceanic crust. As crystallized basalt extruded at a ridge axis cools below Curie points of appropriate iron-titanium oxides, magnetic field directions parallel to the Earth's magnetic field are recorded in those oxides. The orientations of the field in the oceanic crust record preserve a record of directions of the Earth's magnetic field with time. Because the field has reversed directions at irregular intervals throughout its history, the pattern of geomagnetic reversals in the ocean crust can be used as an indicator of age. Likewise, the pattern of reversals together with age measurements of the crust is used to help establish the history of the Earth's magnetic field.

Global System

The mid-ocean ridges of the world are connected and form *the* Ocean Ridge, a single global mid-oceanic ridge system that is part of every ocean, making it the longest mountain range in the world. The continuous mountain range is 65,000 km (40,400 mi) long (several times longer than the Andes, the longest continental mountain range), and the total length of the oceanic ridge system is 80,000 km (49,700 mi) long.

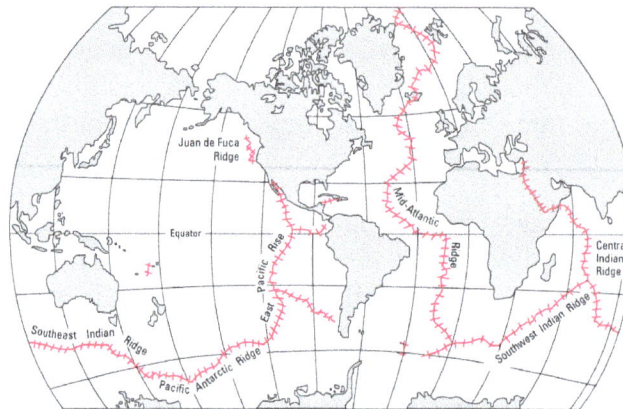

World Distribution of Mid-Oceanic Ridges; USGS

History

Discovery

Mid-ocean ridges are generally submerged deep in the ocean. It was not until the 1950s, when the ocean floor was surveyed in detail, that their full extent became known.

The *Vema*, a ship of the Lamont-Doherty Earth Observatory of Columbia University, traversed the Atlantic Ocean, recording data about the ocean floor from the ocean surface. A team led by Marie Tharp and Bruce Heezen analyzed the data and concluded that there was an enormous mountain chain running up the middle. Scientists named it the "Mid-Atlantic Ridge".

At first, the ridge was thought to be a phenomenon specific to the Atlantic Ocean. However, as surveys of the ocean floor continued around the world, it was discovered that every ocean contains parts of the mid-ocean ridge system. Although the ridge system runs down the middle of the Atlantic Ocean, the ridge is located away from the center of other oceans.

Impact

Alfred Wegener proposed the theory of continental drift in 1912. He stated: "the Mid-Atlantic Ridge ... zone in which the floor of the Atlantic, as it keeps spreading, is continuously tearing open and making space for fresh, relatively fluid and hot sima [rising] from depth." However, Wegener did not pursue this observation in his later works and his theory was dismissed by geologists because there was no mechanism to explain how continents could plow through ocean crust, and the theory became largely forgotten.

Following the discovery of the worldwide extent of the mid-ocean ridge in the 1950s, geologists faced a new task: explaining how such an enormous geological structure could have formed. In the 1960s, geologists discovered and began to propose mechanisms for seafloor spreading. Plate tectonics was a suitable explanation for seafloor spreading, and the acceptance of plate tectonics by the majority of geologists resulted in a major paradigm shift in geological thinking.

It is estimated that 20 volcanic eruptions occur each year along earth's mid-ocean ridges and that every year 2.5 km² (0.97 sq mi) of new seafloor is formed by this process. With a crustal thickness of 1 to 2 km (0.62 to 1.24 mi), this amounts to about 4 km³ (0.96 cu mi) of new ocean crust formed every year.

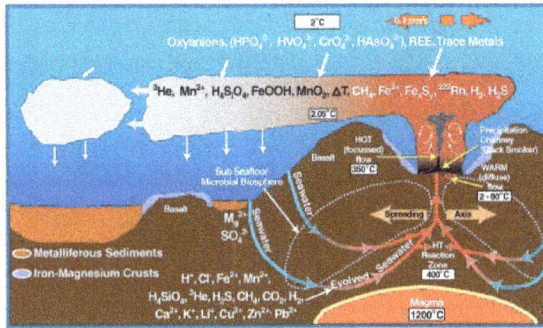

Oceanic ridge and deep sea vent chemistry

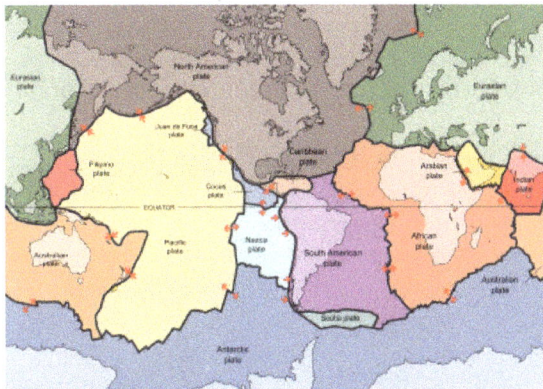

Plates in the crust of the earth, according to the plate tectonics theory

List of Oceanic Ridges

- Aden Ridge

- Cocos Ridge

- Explorer Ridge

- Gorda Ridge

- Juan de Fuca Ridge

- American-Antarctic Ridge

- Chile Rise

- East Pacific Rise

- East Scotia Ridge

- Gakkel Ridge (Mid-Arctic Ridge)

- Nazca Ridge

- Pacific-Antarctic Ridge

- Central Indian Ridge

    o   Carlsberg Ridge

- Southeast Indian Ridge

- Southwest Indian Ridge

- Mid-Atlantic Ridge

    o   Kolbeinsey Ridge (North of Iceland)

    o   Mohns Ridge

    o   Knipovich Ridge (between Greenland and Spitsbergen)

    o   Reykjanes Ridge (South of Iceland)

List of Ancient Oceanic Ridges

- Aegir Ridge

- Alpha Ridge

- Kula-Farallon Ridge

- Pacific-Farallon Ridge

- Pacific-Kula Ridge

- Phoenix Ridge

Various Mid-Ocean Ridge

Mid-Atlantic Ridge

The Mid-Atlantic Ridge (MAR) is a mid-ocean ridge, a divergent tectonic plate or constructive plate boundary located along the floor of the Atlantic Ocean, and part of the longest mountain range in the world. In the North Atlantic, it separates the Eurasian and North American Plates, whereas in the South Atlantic it separates the African and South American Plates. The Ridge extends from a junction with the Gakkel Ridge (Mid-Arctic Ridge) northeast of Greenland southward to the Bouvet Triple Junction in the South Atlantic. Although the Mid-Atlantic Ridge is mostly an underwater feature, portions of it have enough elevation to extend above sea level. The section of the ridge that includes the island of Iceland is also known as the Reykjanes Ridge. The ridge has an average spreading rate of about 2.5 cm per year.

A map of the Mid-Atlantic Ridge

Discovery

A ridge under the Atlantic Ocean was first inferred by Matthew Fontaine Maury in 1850. The ridge was discovered during the expedition of HMS *Challenger* in 1872. A team of scientists on board, led by Charles Wyville Thomson, discovered a large rise in the middle of the Atlantic while investigating the future location for a transatlantic telegraph cable. The existence of such a ridge was confirmed by sonar in 1925 and was found to extend around the Cape of Good Hope into the Indian Ocean by the German Meteor expedition.

In the 1950s, mapping of the Earth's ocean floors by Bruce Heezen, Maurice Ewing, Marie Tharp and others revealed that the Mid-Atlantic Ridge had a strange bathymetry of valleys and ridges, with its central valley being seismologically active and the epicenter of many earthquakes. Ewing, Heezen and Tharp discovered that the ridge is part of a 40,000-km-long essentially continuous system of mid-ocean ridges on the floors of all the Earth's oceans. The discovery of this worldwide ridge system led to the theory of seafloor spreading and general acceptance of Wegener's theory of continental drift and expansion in the modified form of plate tectonics. The ridge is central to the breakup of the hypothetical supercontinent of Pangaea that began some 180 million years ago.

Notable Features

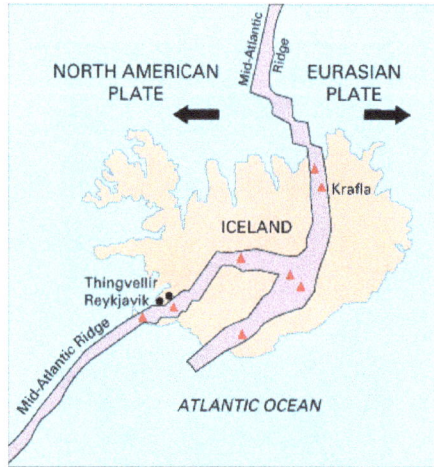

In Iceland the Mid-Atlantic Ridge passes across the Þingvellir
National Park, a popular destination for tourists

The Mid-Atlantic Ridge includes a deep rift valley that runs along the axis of the ridge along nearly its entire length. This rift marks the actual boundary between adjacent tectonic plates, where magma from the mantle reaches the seafloor, erupting as lava and producing new crustal material for the plates.

Near the equator, the Mid-Atlantic Ridge is divided into the North Atlantic Ridge and the South Atlantic Ridge by the Romanche Trench, a narrow submarine trench with a maximum depth of 7,758 m (25,453 ft), one of the deepest locations of the Atlantic Ocean. This trench, however, is not regarded as the boundary between the North and South American Plates, nor the Eurasian and African Plates.

Islands

The islands on or near the Mid-Atlantic Ridge, from north to south, with their respective highest peaks and location, are:

Northern Hemisphere (North Atlantic Ridge):

1. Jan Mayen (Beerenberg, 2277 m (at 71°06′N 08°12′W71.100°N 8.200°W), in the Arctic Ocean

2. Iceland (Hvannadalshnúkur in the Vatnajökull, 2109.6 m (at 64°01′N 16°41′W64.017°N 16.683°W), through which the ridge runs

3. Azores (Ponta do Pico or Pico Alto, on Pico Island, 2351 m, (at 38°28′0″N 28°24′0″W38.46667°N 28.40000°W)

4. Saint Peter and Paul Rocks (Southwest Rock, 22.5 m, at 00°55′08″N 29°20′35″W0.91889°N 29.34306°W)

Southern Hemisphere (South Atlantic Ridge):

1. Ascension Island (The Peak, Green Mountain, 859 m, at 07°59′S 14°25′W7.983°S 14.417°W)

2. Saint Helena (Diana's Peak, 818 m at 15°57′S 5°41′W15.950°S 5.683°W)

3. Tristan da Cunha (Queen Mary's Peak, 2062 m, at 37°05′S 12°17′W37.083°S 12.283°W)

4. Gough Island (Edinburgh Peak, 909 m, at 40°20′S 10°00′W40.333°S 10.000°W)

5. Bouvet Island (Olavtoppen, 780 m, at 54°24′S 03°21′E54.400°S 3.350°E)

Geology

Basaltic rocks of the MAR collected by the Hercules ROV during the 2005 Lost City Expedition.

The ridge sits atop a geologic feature known as the Mid-Atlantic Rise, which is a progressive bulge that runs the length of the Atlantic Ocean, with the ridge resting on the highest point of this linear bulge. This bulge is thought to be caused by upward convective forces in the asthenosphere pushing the oceanic crust and lithosphere. This divergent boundary first formed in the Triassic period, when a series of three-armed grabens coalesced on the supercontinent Pangaea to form the ridge. Usually, only two arms of any given three-armed graben become part of a divergent plate boundary. The failed arms are called *aulacogens*, and the aulacogens of the Mid-Atlantic Ridge eventually became many of the large river valleys seen along the Americas and Africa (including the Mississippi River, Amazon River and Niger River). The Fundy Basin on the Atlantic coast of North America between New Brunswick and Nova Scotia in Canada is evidence of the ancestral Mid-Atlantic Ridge.

South American–Antarctic Ridge

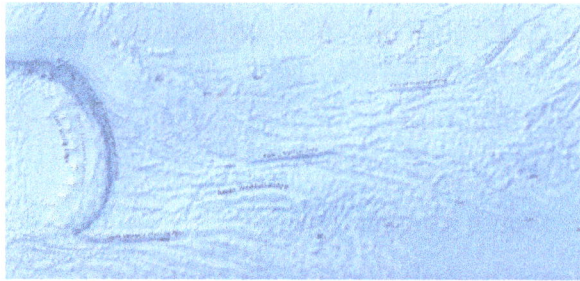

Bathymetric map of the South American-Antarctic Ridge

The South American–Antarctic Ridge (SAAR or AAR) is the tectonic spreading center between the South American Plate and the Antarctic Plate. It runs along the sea-floor from the Bouvet Triple Junction in the South Atlantic Ocean south-westward to a major transform fault boundary east of the South Sandwich Islands. Near the Bouvet Triple Junction the spreading half rate is 9 mm/a (0.011 in/Ms), which is slow, and the SAAR has the rough topography characteristic of slow-spreading ridges.

Geologic Setting

The boundary between the South American and Antarctic plates can be divided into three parts of which the SAAR forms the eastern third:

The first stretches from the Chile Triple Junction in the Chile Trench at 46°S to the western Straits of Magellan at 52°S. Since 15 Ma, the oceanic crust of the Antarctic plate is being slowly subducted (20–24 mm/a (0.025–0.030 in/Ms)) under South America along this trench which is currently extending northward. In the central part, between the Straits of Magellan and the South Sandwich Trench, the two large continental plates are separated by the Scotia Plate and a number of smaller plates east of it. During the past 40 Ma (or since the opening of the Drake Passage) the South Sandwich Trench has been migrating eastward due to the evolution of a back-arc basin, effectively consuming the SAAR.

The third eastern part, i.e. the 'SAAR proper', has two long and several shorter transform faults separating short north-to-south-directed ridge crests. The motion in the SAAR is currently c. 20 mm/a (0.025 in/Ms) westward but it was originally closer to north-south. It can be inferred, based on fracture zone topography and magnetic anomalies in the Weddell Sea, that this change in direction occurred during the Cretaceous and Cenozoic. The western part of the SAAR is dominated by the earthquake-intensive South Sandwich island arc, fore-arc, and trench. East of these structures the SAAR is composed of a series of north-south-oriented ridge crests, median valleys, and east-west-oriented transform faults. The latter are mostly short (less than 100 km) with the exceptions of much longer (east to west) Conrad (200 km), Bullard, and South Sandwich (both over 400 km) fracture zones. The topography of the SAAR is extreme, with valleys reaching 1.5–2 km (0.93–1.24 mi) deeper than adjacent ridges in average and maximum depth exceeding 3 km. The SAAR is more shallow near the Bouvet Triple Junction.

North of the SAAR the South Sandwich Plate consumes the South American Plate at a rate of 65.8 mm/yr driven by back-arc extension. This fast subduction has broken off the southern part of the South American Plate between the north-eastern end of the South Sandwich Arc and the Mid-Atlantic Ridge, leaving a separate microplate called 'Sur' (Spanish for 'South') north of the SAAR. The southern part of this Sur microplate has probably also been broken off and is subducting independently under the South Sandwich Arc.

Tectonic Evolution

The break-up of Gondwana began in the Mid- to Late Jurassic in what is today the Mozambique Basin east of Africa, whereas the South American and African plates started to brake apart during the Early Cretaceous. Between these events neither the Mid-Atlantic Ridge nor the Bouvet Triple Junction existed and the SAAR formed a continuous ridge together with the Southwest Indian Ridge.

Around 106 Ma the eastern end of the Falkland Plateau separated from the Agulhas Bank, opening the South Atlantic which, however, remained an enclosed basin north of the Falkland Plateau until 85-83 Ma. Around 97 Ma the Northeast Georgia Rise (today north of the Scotia Plate) and the Maud Rise (off Antarctica) were located next to the Agulhas Plateau (south of South Africa) where the Bouvet hotspot formed the Southern Ocean Large Igneous Province (112–93 Ma).

The Bouvet Triple Junction, today considered an R-F-F (ridge-fault-fault) type triple junction, was an R-R-R type before anomaly 28 (c. 64 Ma), which means that before the Scotia Plate started to develop in the Mid-Tertiary, only ridges and transform faults separated Africa, Antarctica, and South America. The north-south motion of Antarctica relative to Africa and South America before anomaly 28 changed to a slow east-west clockwise motion around 60 Ma, an abrupt change coincident with change in triple junction configuration.

During the Late Paleocene-Early Eocene Antarctica and South America separated at a rate of only 0.3 cm/yr. The brief opening of small extensional basins south of Tierra del Fuego initiated the opening of the Drake Passage around 49 Ma after which the spreading rate increased eight-fold to 2.4 cm/yr. As the Antarctic—South American plate motion changed from north to north-west during this period, oceanic crust in the north-west Weddell Sea started to subduct on the eastern side of the Drake land bridge — the eastward migration of the South Sandwich Trench had begun.

Central Indian Ridge

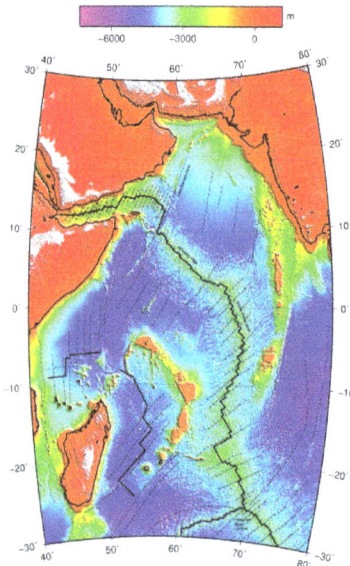

Bathymetric map of Central Indian Ridge.

The Central Indian Ridge (CIR) is a north-south-trending mid-ocean ridge in the western Indian Ocean.

Geological Setting

The morphology of the CIR is characteristic of slow to intermediate ridges. The axial valley is 500–1000 m deep; 50–100 km-long ridge segments are separated by 30 km-long transform faults and 10 km-long non-transform discontinuities. Melt supply comes from axial volcanic ridges that are 15 km-long, 1–2 km wide, and reaches 100–200 m above the axial floor.

With a spreading rate of 30 mm/yr near the Equator and 49 mm/yr near the Rodrigues Triple Junction (RTJ) at its southern end, the CIR is an intermediately fast spreading ridge characterised by moderate obliquity and few large offsets, the obvious exception being the almost 300 km-long Marie Celeste Fracture Zone at 18°S. Between 21°S and the Marie Celeste Fracture Zone (18°S) the CIR deviates westward. Along this section the larger offsets switch from right-lateral to left-lateral but return to right-lateral north of 18°S.

Otherwise, the southern section (RTJ-Argo Fracture Zone, 25°S-13°S) of the CIR is near-orthogonal relative to the spreading direction. North of the Argo FZ it is highly oblique and dominated by numerous small ridge segments. The northern section of the CIR, including the Carlsberg Ridge, trends NNW and lacks fracture zones. The axial depth of the CIR increases from 3200 m at 20°S to 4000 m at the RTJ.

Boundaries

The CIR is traditionally said to separate the African Plate from the Indo-Australian Plate. Likewise, the Owen Fracture Zone in the northern end of the CIR is traditionally said to separate the Indian-Australian plate from the Arabian Plate. Movements in the Owen Fracture Zone are, however, negligible and Arabia and India are moving as a single plate. This plate, in turn, is separated from the Australian Plate by a diffuse boundary, the India–Capricorn boundary, which stretches east from the CIR near Chagos Bank to the Ninetyeast Ridge and north along the Ninetyeast Ridge to the northern end of the Sunda Trench. This diffuse boundary was probably initiated in the Late Miocene and is probably related to opening of Gulf of Aden and the uplift of the Himalayas.

Tectonic History and Hotspot Interaction

The CIR was opened during the separation of the Mascarene Plateau and the Chagos-Laccadive Ridge about 38 Ma, both of which are the products of the Réunion hotspot, the only hotspot known to have interacted with the CIR. Now located 1100 km from the CIR, hotspot crossed the CIR near 18-20°S, from the Indian to the African plate, at 47 Ma. The Réunion hotspot track includes the Chagos-Laccadive Ridge on the Indian Plate which leads to the Indian west-coast where the newborn hotspot produced the Deccan Traps in north-west India at 66 Ma.

The only above-water structure near the CIR is the Rodrigues Island, the top of the enigmatic Rodrigues Ridge between Mauritius and the CIR. The Rodrigues Ridge reaches the CIR at 19°S via a series of en echelon ridges known as the Three Magi. Volcanic rocks from the Rodrigues Island are, however, similar to 1,58-1,30 Ma-old rocks from Réunion and Mauritius and the Rodrigues Ridge can't therefore have originated on the CIR leaving the Réunion hotspot the most likely candidate.

Seafloor Spreading

Seafloor spreading is a process that occurs at mid-ocean ridges, where new oceanic crust is formed through volcanic activity and then gradually moves away from the ridge.

Seafloor spreading helps explain continental drift in the theory of plate tectonics. When oceanic plates diverge, tensional stress causes fractures to occur in the lithosphere. Basaltic magma rises up the fractures and cools on the ocean floor to form new sea floor. Older rocks will be found farther away from the spreading zone while younger rocks will be found nearer to the spreading zone.

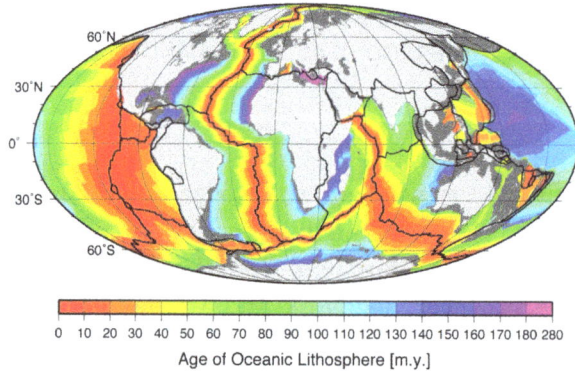

Age of oceanic lithosphere; youngest (red) is along spreading centers.

Earlier theories (e.g. by Alfred Wegener and Alexander du Toit) of continental drift were that continents "ploughed" through the sea. The idea that the seafloor itself moves (and carries the continents with it) as it expands from a central axis was proposed by Harry Hess from Princeton University in the 1960s. The theory is well accepted now, and the phenomenon is known to be caused by convection currents in the asthenosphere, which is the plastic, relatively weak part of the upper mantle.

Incipient Spreading

In the general case, sea floor spreading starts as a rift in a continental land mass, similar to the Red Sea-East Africa Rift System today. The process starts with heating at the base of the continental crust which causes it to become more plastic and less dense. Because less dense objects rise in relation to denser objects, the area being heated becomes a broad dome. As the crust bows upward, fractures occur that gradually grow into rifts. The typical rift system consists of three rift arms at approximately 120 degree angles. These areas are named triple junctions and can be found in several places across the world today. The separated margins of the continents evolve to form passive margins. Hess' theory was that new seafloor is formed when magma is forced upward toward the surface at a mid-ocean ridge.

If spreading continues past the incipient stage described above, two of the rift arms will open while the third arm stops opening and becomes a 'failed rift'. As the two active rifts continue to open, eventually the continental crust is attenuated as far as it will stretch. At this point basaltic oceanic crust begins to form between the separating continental fragments. When one of the rifts opens into the existing ocean, the rift system is flooded with seawater and becomes a new sea. The Red Sea is an example of a new arm of the sea. The East African rift was thought to be a "failed" arm that was opening somewhat more slowly than the other two arms, but in 2005 the Ethiopian Afar Geophysical Lithospheric Experiment reported that in the Afar region last September, a 60 km fissure opened as wide as eight meters. During this period of initial flooding the new sea is sensitive to changes in climate and eustasy. As a result, the new sea will evaporate (partially or completely) several times before the elevation of the rift valley has been lowered to the point that the sea becomes stable. During this period of evaporation large evaporite deposits will be made in the rift valley. Later these deposits have the potential to become hydrocarbon seals and are of particular interest to petroleum geologists.

Sea floor spreading can stop during the process, but if it continues to the point that the continent is completely severed, then a new ocean basin is created. The Red Sea has not yet completely split Arabia from Africa, but a similar feature can be found on the other side of Africa that has broken completely free. South America once fit into the area of the Niger Delta. The Niger River has formed in the failed rift arm of the triple junction.

Continued Spreading and Subduction

As new seafloor forms and spreads apart from the mid-ocean ridge it slowly cools over time. Older seafloor is therefore colder than new seafloor, and older oceanic basins deeper than new oceanic basins due to isostasy. If the diameter of the earth remains relatively constant despite the production of new crust, a mechanism must exist by which crust is also destroyed. The destruction of oceanic crust occurs at subduction zones where oceanic crust is forced under either continental crust or oceanic crust. Today, the Atlantic basin is actively spreading at the Mid-Atlantic Ridge. Only a small portion of the oceanic crust produced in the Atlantic is subducted. However, the plates making up the Pacific Ocean are experiencing subduction along many of their boundaries which causes the volcanic activity in what has been termed the Ring of Fire of the Pacific Ocean. The Pacific is also home to one of the world's most active spreading centres (the East Pacific Rise) with spreading rates of up to 13 cm/yr. The Mid-Atlantic Ridge is a "textbook" slow-spreading centre, while the East Pacific Rise is used as an example of fast spreading. The differences in spreading rates affect not only the geometries of the ridges but also the geochemistry of the basalts that are produced.

Since the new oceanic basins are shallower than the old oceanic basins, the total capacity of the world's ocean basins decreases during times of active sea floor spreading. During the opening of the Atlantic Ocean, sea level was so high that a Western Interior Seaway formed across North America from the Gulf of Mexico to the Arctic Ocean.

Debate and Search for Mechanism

At the Mid-Atlantic Ridge (and in other areas), material from the upper mantle rises through the faults between oceanic plates to form new crust as the plates move away from each other, a phenomenon first observed as continental drift. When Alfred Wegener first presented a hypothesis of continental drift in 1912, he suggested that continents ploughed through the ocean crust. This was impossible: oceanic crust is both more dense and more rigid than continental crust. Accordingly, Wegener's theory wasn't taken very seriously, especially in the United States.

Since then, it has been shown that the motion of the continents is linked to seafloor spreading. In the 1960s, the past record of geomagnetic reversals was noticed by observing the magnetic stripe "anomalies" on the ocean floor. This results in broadly evident "stripes" from which the past magnetic field polarity can be inferred by looking at the data gathered from simply towing a magnetometer on the sea surface or from an aircraft. The stripes on one side of the mid-ocean ridge were the mirror image of those on the other side. The seafloor must have originated on the Earth's great fiery welts, like the Mid-Atlantic Ridge and the East Pacific Rise.

The driver for seafloor spreading in plates with active margins is the weight of the cool, dense, subducting slabs that pull them along. The magmatism at the ridge is considered to be "passive upswelling", which is caused by the plates being pulled apart under the weight of their own slabs. This can be thought of as analogous to a rug on a table with little friction: when part of the rug is off of the table, its weight pulls the rest of the rug down with it.

Sea Floor Global Topography: Half-space Model

To first approximation, sea floor global topography in areas without significant subduction can be estimated by the half-space model. In this model, the seabed height is determined by the oceanic lithosphere temperature, due to thermal expansion. Oceanic lithosphere is continuously formed at a constant rate at the mid-ocean ridges. The source of the lithosphere has a half-plane shape (x = 0, z < 0) and a constant temperature T_1. Due to its continuous creation, the lithosphere at x > 0 is moving away from the ridge at a constant velocity v, which is assumed large compared to other typical scales in the problem. The temperature at the upper boundary of the lithosphere (z=0) is a constant $T_0 = 0$. Thus at x = 0 the temperature is the Heaviside step function $T_1 \cdot \Theta(-z)$. Finally, we assume the system is at a quasi-steady state, so that the temperature distribution is constant in time, i.e. T=T(x,z).

By calculating in the frame of reference of the moving lithosphere (velocity v), which have spatial coordinate x' = x-vt, we may write T = T(x',z,t) and use the heat equation:

$$\frac{\partial T}{\partial t} = \kappa \nabla^2 T = \kappa \frac{\partial^2 T}{\partial^2 z} + \kappa \frac{\partial^2 T}{\partial^2 x'}$$ where κ is the thermal diffusivity of the mantle lithosphere.

Since T depends on x' and t only through the combination $x = x' + vt$, we have:

$$\frac{\partial T}{\partial x'} = \frac{1}{v} \cdot \frac{\partial T}{\partial t}$$

$$\text{Thus: } \frac{\partial T}{\partial t} = \kappa \nabla^2 T = \kappa \frac{\partial^2 T}{\partial^2 z} + \frac{\kappa}{v^2} \frac{\partial^2 T}{\partial^2 t}$$

We now use the assumption that v is large compared to other scales in the problem; we therefore neglect the last term in the equation, and get a 1-dimensional diffusion equation:

$\frac{\partial T}{\partial t} = \kappa \frac{\partial^2 T}{\partial^2 z}$ with the initial conditions $T(t = 0) = T_1 \cdot \Theta(-z)$.

The solution for $z \leq 0$ is given by the error function erf :

$$T(x', z, t) = T_1 \cdot \text{erf}\left(\frac{z}{2\sqrt{\kappa t}}\right).$$

Due to the large velocity, the temperature dependence on the horizontal direction is negligible, and the height at time t (i.e. of sea floor of age t) can be calculated by integrating the thermal expansion over z.

$$h(t) = h_0 + \alpha_{\text{eff}} \int_0^\infty [T(z) - T_1] dz = h_0 - \frac{2}{\sqrt{\pi}} \alpha_{\text{eff}} T_1 \sqrt{\kappa t}$$

Note that the assumption the v is relatively large is equivalently to the assumption that the thermal diffusivity κ is small compared to L^2/T, where L is the acean width (from mid-ocean ridges to continental shelf) and T is its age.

The effective thermal expansion coefficient α_{eff} is different from the usual thermal expansion coefficient α due to isostasic effect of the change in water column height above the lithosphere as it expands or retracts. Both coefficients are related by:

$$\alpha_{\text{eff}} = \alpha \cdot \frac{\rho}{\rho - \rho_w}$$

where $\rho \sim 3.3 g/cm^3$ is the rock density and $\rho_0 = 1 g/cm^3$ is the density of water.

By substituting the parameters by their rough estimates: $\kappa \sim 8 \cdot 10^{-7} \, m^2/s$, $\alpha \sim 4 \cdot 10^{-5}$ °C⁻¹ and $T_1 \sim 1220$ °C (for the Atlantic and Indian oceans) or ~1120 °C (for the eastern Pacific), we have:

$$h(t) \sim h_0 - 350\sqrt{t}$$

for the eastern Pacific Ocean, and:

$$h(t) \sim h_0 - 390\sqrt{t}$$

for the Atlantic and Indian Ocean, where the height is in meters and time is in millions of years. To get the dependence on x, one must substitute t = x/v ~ Tx/L, where L is the distance between the ridge to the continental shelf (roughly half the ocean width), and T is the ocean age.

Galápagos Hotspot

The Galápagos hotspot is a volcanic hotspot in the East Pacific Ocean responsible for the creation of the Galapagos Islands as well as three major aseismic ridge systems, Carnegie, Cocos and Malpelso which are on two tectonic plates. The hotspot is located near the Equator on the Nazca Plate not far from the divergent plate boundary with the Cocos Plate. The tectonic setting of the hotspot is complicated by the Galapagos Triple Junction of the Nazca and Cocos plates with the Pacific Plate. The movement of the plates over the hotspot is determined not solely by the spreading along the ridge but also by the relative motion between the Pacific Plate and the Cocos and Nazca Plates.

The hotspot is believed to be over 20 million years old and in that time there has been interaction between the hotspot, both of these plates, and the divergent plate boundary, at the Galapagos Spreading Centre. Lavas from the hotspot do not exhibit the homogeneous nature of many hotspots; instead there is evidence of four major reservoirs feeding the hotspot. These mix to varying degrees at different locations on the archipelago and also within the Galapagos Spreading Centre.

Hotspot Theory

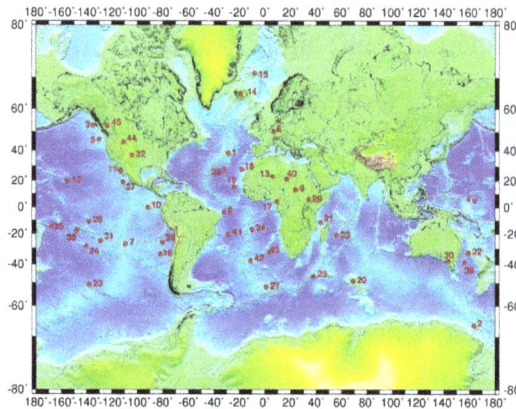

The Galápagos hotspot is marked 10 on the map.

In 1963, Canadian geophysicist J. Tuzo Wilson proposed the "hotspot" theory to explain why although most earthquake and volcanic activity occurs at plate boundaries, some occurs far from plate boundaries. The theory claimed that small, long-lasting, exceptionally "hot" areas of magma are located under certain points on Earth. These places, dubbed "hotspots," provide localized heat and energy systems (thermal plumes) that sustain long-lasting volcanic activity on the surface. This volcanism builds up seamounts that eventually rise above the ocean current, forming volcanic islands. As the islands slowly moved away from the hotspot, by the motion of sliding plates as described by the theory of plate tectonics, the magma supply is cut, and the volcano goes dormant. Meanwhile, the process repeats all over again, this time forming a new island, on and on until the hotspot collapses. The theory was developed to explain the Hawaiian-Emperor seamount chain, where historic islands could be traced to the northwest in the direction that the Pacific Plate is moving. The early theory put these fixed sources of heat for the plumes deep within the Earth; however, recent research has led scientists to believe that hotspots are actually dynamic, and able to move on their own accord.

Tectonic Setting

The Galapagos hotspot has a very complicated tectonic setting. It is located very close to the spreading ridge between the Cocos and Nazca plates; the hotspot interacts with both plates and the spreading ridge over the last twenty million years as the relative location of the hotspot in relation to the plates has varied. Based on similar seismic velocity gradients of the lavas of the Carnegie, Cocos and Malpelos Ridges there is evidence that the hotspot activity has been the result of a single long mantle melt rather than multiple periods of activity and dormancy.

In Hawaii the evidence suggests that each volcano has a distinct period of activity as the hotspot moves under that portion of the Pacific plate before becoming dormant and then extinct and eroding under the ocean. This does not appear to be the case in the Galapagos, instead there is evidence of concurrent volcanism over a wide area. Nearly all Galapagos Islands show volcanism in the recent geological past, not just at the current location of the hotspot at Fernandina. The list below gives the last eruption dates for the Galapagos volcanoes, ordered from West to East.

The movement of the Nazca and Cocos plates have been tracked. The Nazca plate moves at 90 degrees at a rate of 58 ± 2 km per million years. The Cocos Plate moves at 41 degrees at a rate of 83 ± 3 km per million years. The location of the hotspot over time is recorded in the oceanic plate as the Carnegie and Cocos Ridges.

The Carnegie Ridge is on the Nazca plate is 600 km (373 mi) long and up to 300 km (186 mi) wide. It is orientated parallel to the plate movement, and its eastern end is approximately 20 million years old. There is a prominent saddle in the ridge at 86 degrees West where the height drops much closer to the surrounding ocean floor. The Malpelo Ridge, which is 300 km (186 mi) long was once believed to be part of the Carnegie Ridge.

The Cocos Ridge is 1000 km long located on the Cocos plate and is orientated parallel to the plates motion from the 91 degree west transform fault at the Galapagos Spreading Centre towards the Panamanian cost. The north eastern end of the ridge dates from about 13-14.5 million years ago. However the Cocos Islands to the northern end of the ridge are only 2 million years old, and were therefore created at a time well after the ridge had moved away from the hotspot.

The current model for the interaction of the hotspot and the spreading centre between the Cocos and Nazca plates attempts to explain the ridges on both plates; the split between the Carnegie and Malpelo Ridge and subsequent volcanic activity away from the hotspot. There have been eight major phases in the last 20 million years.

(1)   19.5 million years - 14.5 million years ago: the hotspot was located on the Nazca plate, forming a combined Carnegie and Malpelo Ridge. The type of lava erupted was a mix of plume material and depleted upper mantle, similar to the type of lava found in the central Galapagos islands at the current time.

(2)   From 14.5 million years to 12.5 million years ago: the Galapagos Spreading Centre moved south and the ridge overlay the southern edge of the hotspot. Less material is erupted over the Nazca plate resulted in the saddle being formed in the Carnegie Ridge. The movement of the location of the Galapagos Spreading Centre starts to rift the Malpelo Ridge away from the Carnegie Ridge. The majority of the hotspot lavas are created on the Cocos plate resulting in the formation of the Cocos Ridge. The lavas formed here are similar to the types erupted on the western shield volcanoes of the Galapagos, which are predominantly plume.

(3)   12 million years to 11 million years: The Galapagos hotspot is centred under the Galapagos Spreading Centre. plume-type lavas are now abundant on the Cocos Ridge.

(4)   9.5 million years ago: the rifting between the Carnegie and Malpelo Ridges ends.

(5)   5.2 million years ago to 3.5 million years ago: the Galapagos Spreading Centre has another ridge jump, moving northwards with the plume now erupting on the Nazca plate, similar to the present orientation.

(6)   3.5 million to 2 million years ago: A short-lived east–west trending spreading centre is formed north of the Galapagos Spreading Centre. This new rift fails but leads to post abandonment volcanic activity and the subsequent formation of the Cocos Islands and surrounding seamounts. Around the hotspot plume lavas predominate.

(7)   2.6 million years ago: a major transform fault occurs north of the Galapagos hotspot. This results in widespread volcanism in the northern Galapagos along the Wolf Darwin Lineament and around Genovesa Island.

(8)   Present : The Galapagos hotspot is south of the spreading centre and there is geochemical zonation of the plume.

Chemical Structure of the Galapagos Lavas

Analysis of the radioactive isotopes of the lavas on the islands of the Galapagos archipelago and on the Carnegie Ridge shows that there are four major reservoirs of magma that mix in varying combinations to form the volcanic province.

The four types are: PLUME – this is magma associated with the plume itself and is similar to magmas from other ocean islands within the Pacific. It has the characteristics of intermediate Strontium (Sr), Neodymium (Nd) and Lead (Pb) ratios. The PLUME lavas are found predominately in the west of the islands, around Ferdinandina and Isabela Islands, which is near to the current position of the hotspot. The PLUME lavas erupted on Fernandina and Isabela are relatively cool. Analysis shows that they are as much as 100 degrees Celsius cooler than those in Hawaii. The cause of this is not fully understood but may be due to cooling in the lithosphere or being relatively cool at formation in the mantle. They are then found in lower quantities in a horseshoe pattern north and south of the central islands mixing with the other reservoirs as it progresses east. PLUME lavas are also found in the lavas from the Galapagos Spreading Centre due to convection and mixing of all of these lavas. In the upper mantle convection currents bring in mantle material at shallow angles from the south of the Galapagos Spreading Centre. These convection current will draw in some PLUME type magma to the spreading centre where it is then erupted.

DGM – (Depleted Galapagos Mantle), this has similar characteristics to ocean ridge basalts throughout the Pacific and the Galapagos Spreading Centre. Partial melting of the upper mantle as a result of the spreading centre will leave mantle material depleted in some compounds. It has low Sr and Pb isotope ratios and high Nd ratios. DGM is found in the central islands of the Galapagos such as Santiago, Santa Cruz, San Cristobal and Santa Fe. It fills in the centre of the horseshoe formed by the PLUME lavas to the west, north and south.

FLO – (Floreana), characteristic of that islands lavas. It is thought that this reservoir came from subducted ocean crust that has been entrained by the mantle plume. It has enriched Sr and Pb ratios and is enriched with trace elements. FLO is associated principally with the island of Floreana and shows up on the mixing of lavas within the Galapagos along the southern side archipelago and is diluted to the east and north of there.

WD – (Wolf Darwin) is unique in the Pacific and resembles material from an Indian Ocean Ridge system. It is found on the Wolf and Darwin Islands and the seamounts that connect them along the Wolf Darwin Lineament. It has a unique Pb ratio. WD is located along the northern side of the archipelago and dilutes to the east and south.

Oceanic Trench

The oceanic trenches are hemispheric-scale long but narrow topographic depressions of the sea floor. They are also the deepest parts of the ocean floor. Oceanic trenches are a distinctive morphological feature of convergent plate boundaries, along which lithospheric plates move towards each other at rates that vary from a few mm to over ten cm per year. A trench marks the position at which the flexed, subducting slab begins to descend beneath another lithospheric slab. Trenches are generally parallel to a volcanic island arc, and about 200 km (120 mi) from a volcanic arc. Oceanic trenches typically extend 3 to 4 km (1.9 to 2.5 mi) below the level of the surrounding oceanic floor. The greatest ocean depth to be sounded is in the Challenger Deep of the Mariana Trench, at a depth of 11,034 m (36,201 ft) below sea level. Oceanic lithosphere moves into trenches at a global rate of about 3 km^2/yr.

Geographic Distribution

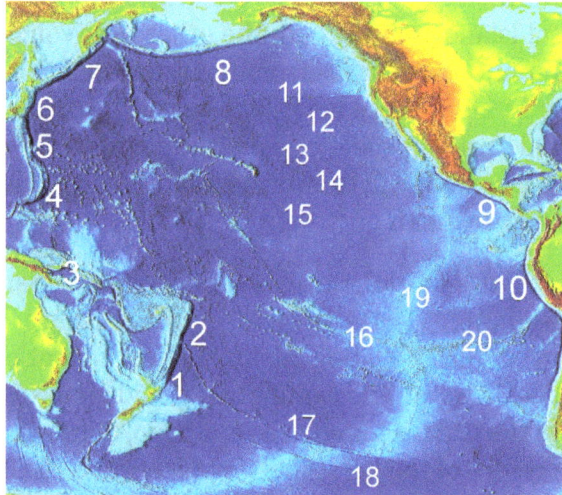

Major Pacific trenches (1-10) and fracture zones (11-20): 1. Kermadec 2. Tonga 3. Bougainville 4. Mariana 5. Izu-Ogasawara 6. Japan 7. Kuril–Kamchatka 8. Aleutian 9. Middle America 10. Peru-Chile 11. Mendocino 12. Murray 13. Molokai 14. Clarion 15. Clipperton 16. Challenger 17. Eltanin 18. Udintsev 19. East Pacific Rise (S-shaped) 20. Nazca Ridge

There are about 50,000 km (31,000 mi) of convergent plate margins, mostly around the Pacific Ocean—the reason for the reference "Pacific-type" margin—but they are also in the eastern Indian Ocean, with relatively short convergent margin segments in the Atlantic Ocean and in the Mediterranean Sea. Globally, there are over 50 major ocean trenches covering an area of 1.9 million km2 or about 0.5% of the oceans. Trenches that are partially infilled are known as "troughs" and sometimes they are completely buried and lack bathymetric expression, but the fundamental structures that these represent mean that the great name should also be applied here. This applies to Cascadia, Makran, southern Lesser Antilles, and Calabrian trenches. Trenches along with volcanic arcs and zones of earthquakes that dip under the volcanic arc as deeply as 700 km (430 mi) are diagnostic of convergent plate boundaries and their deeper manifestations, subduction zones. Trenches are related to but distinguished from continental collision zones (like that between India and Asia to form the Himalaya), where continental crust enters the subduction zone. When buoyant continental crust enters a trench, subduction eventually stops and the convergent plate margin becomes a collision zone. Features analogous to trenches are associated with collisions zones; these are sediment-filled foredeeps referred to as peripheral foreland basins, such as that which the Ganges River and Tigris-Euphrates rivers flow along.

History of the Term "Trench"

Trenches were not clearly defined until the late 1940s and 1950s. The bathymetry of the ocean was of no real interest until the late 19th and early 20th centuries, with the initial laying of Transatlantic telegraph cables on the seafloor between the continents. Even then the elongated bathymetric expression of trenches was not recognized until well into the 20th century. The term "trench" does not appear in Murray and Hjort's (1912) classic oceanography book. Instead they applied the term "deep" for the deepest parts of the ocean, such as Challenger Deep. Experiences from World War I battlefields emblazoned the concept of the trench warfare as an elongate depression defining an important boundary, so it was no surprise that the term "trench" was used to describe natural features in the early 1920s. The term was first used in a geologic context by Scofield two years after the war ended to describe a structurally controlled depression in the Rocky Mountains. Johnstone, in his 1923 textbook *An Introduction to Oceanography*, first used the term in its modern sense for any marked, elongate depression of the sea bottom.

During the 1920s and 1930s, Felix Andries Vening Meinesz developed a unique gravimeter that could measure gravity in the stable environment of a submarine and used it to measure gravity over trenches. His measurements revealed that trenches are sites of downwelling in the solid Earth. The concept of downwelling at trenches was characterized by Griggs in 1939 as the tectogene hypothesis, for which he developed an analogue model using a pair of rotating drums. World War II in the Pacific led to great improvements of bathymetry in especially the western and northern Pacific, and the linear nature of these deeps became clear. The rapid growth of deep sea research efforts, especially the widespread use of echosounders in the 1950s and 1960s confirmed the morphological utility of the term. The important trenches were identified, sampled, and their greatest depths sonically plumbed. The heroic phase of trench exploration culminated in the 1960 descent of the Bathyscaphe *Trieste*, which set an unbeatable world record by diving to the bottom of the Challenger Deep. Following Robert S. Dietz' and Harry Hess' articulation of the seafloor spreading hypothesis in the early 1960s and the plate tectonic revolution in the late 1960s the term "trench" has been redefined with plate tectonic as well as bathymetric connotations.

Trench Rollback

Although trenches would seem to be positionally stable over time, it is hypothesized that some trenches, particularly those associated with subduction zones where two oceanic plates converge, they move backward into the plate which is subducting, akin to a backward-moving wave. This has been termed trench rollback or hinge retreat (also hinge rollback). This is one explanation for the existence of back-arc basins.

Slab rollback is a process which occurs during the subduction of two tectonic plates resulting in the seaward motion of the trench. Forces acting perpendicular to the slab (portion of the subducting plate within the mantle) at depth are responsible for the backward migration of the slab in the mantle and ultimately the movement of the hinge and trench at the surface. The driving force for rollback is the negative buoyancy of the slab with respect to the underlying mantle as well as the geometry of the slab. Back-arc basins are often associated with slab rollback due to extension in the overriding plate as a response to the subsequent subhorizontal mantle flow from the displacement of the slab at depth.

Processes Involved

Several forces are involved in the processes of slab rollback. Two forces acting against each other at the interface of the two subducting plates exert forces against one another. The subducting plate exerts a bending force (FPB) which is the pressure supplied during subduction, while the overriding plate exerts a force against the subducting plate (FTS). The slab pull force (FSP) is caused by the negative buoyancy of the plate driving the plate to greater depths. The resisisting force from the surrounding mantle opposes the slab pull forces. Interactions with the 660-km discontinuity will cause a deflection due to the buoyancy at the phase transition (F660). The unique interplay of these forces is what generates slab rollback. When the deep slab section obstructs the down-going motion of the shallow slab section, slab rollback will occur. The subducting slab undergoes backward sinking due to the negative buoyancy forces causing a retrogradation of the trench hinge along the surface. Upwelling of the mantle around the slab can create favorable conditions for the formation of a back-arc basin.

Seismic tomography provides evidence for slab rollback. Results demonstrate high temperature anomalies within the mantle suggesting subducted material is present in the mantle. Ophiolites are viewed as evidence for such mechanisms as high pressure and temperature rocks are rapidly brought to the surface through the processes of slab rollback which provides space for the exhumation of ophiolites.

Slab rollback is not always a continuous process suggesting an episodic nature. The episodic nature of the rollback is explained by a change in the density of the subducting plate, such as the arrival of buoyant lithosphere (a continent, arc, ridge, or plateau), a change in the subduction dynamics, or a change in the plate kinematics. The age of the subducting plates does not have any effect on slab rollback. Nearby continental collisions have an effect on slab rollback. Continental collisions induce mantle flow and extrusion of mantle material which results in stretching and arc-trench rollback. In the area of the Southeast Pacific, there have been several rollback events resulting in the formation of numerous back-arc basins.

Mantle Interactions

Interactions with the mantle discontinuities play a significant role in slab rollback. Stagnation at the 660-km discontinuity causes retrograde slab motion due to the suction forces acting at the surface. Slab rollback induces mantle return flow which causes extension from the shear stresses at the base of the overriding plate. As slab rollback velocities increase, circular mantle flow velocities also increase, accelerating extension rates. Extension rates are altered when the slab interacts with the discontinuities within the mantle at 410 km and 660 km depth. Slabs can either penetrate directly into the lower mantle, or can be retarded due to the phase transition at 660 km depth creating a difference in buoyancy. An increase in retrograde trench migration (slab rollback) (2–4 cm/yr) is a result of flattened slabs at the 660-km discontinuity where the slab does not penetrate into the lower mantle. This is the case for the Japan, Java and Izu-Bonin trenches. These flattened slabs are only temporarily arrested in the transition zone. The subsequent displacement into the lower mantle is caused by slab pull forces, or the destabilization of the slab from warming and broadening due to thermal diffusion. Slabs that penetrate directly into the lower mantle result in slower slab rollback rates (~1–3 cm/yr) such as the Mariana arc, Tonga arcs.

Morphologic Expression

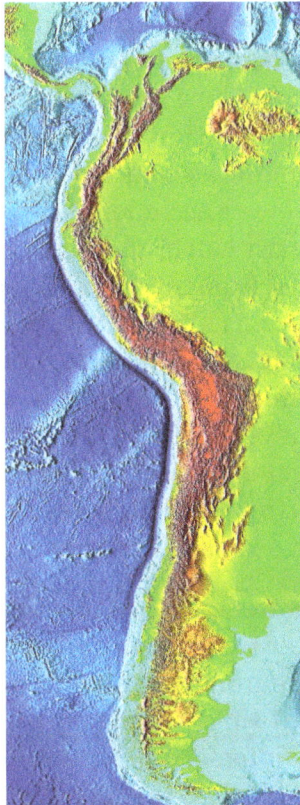

The Peru–Chile Trench

Trenches are centerpieces of the distinctive physiography of a convergent plate margin. Transects across trenches yield asymmetric profiles, with relatively gentle (~5°) outer (seaward) slope and a steeper (~10–16°) inner (landward) slope. This asymmetry is due to the fact that the outer slope is defined by the top of the downgoing plate, which must bend as it starts its descent. The great thickness of the lithosphere requires that this bending be gentle. As the subducting plate approaches the trench, it is first bent upwards to form the outer trench swell, then descends to form the outer trench slope. The outer trench slope is disrupted by a set of subparallel normal faults which staircase the seafloor down to the trench. The plate boundary is defined by the trench axis itself. Beneath the inner trench wall, the two plates slide past each other along the subduction decollement, the seafloor intersection of which defines the trench location. The overriding plate contains volcanic arc (generally) and a forearc. The volcanic arc is caused by physical and chemical interactions between the subducted plate at depth and asthenospheric mantle associated with the overriding plate. The forearc lies between the trench and the volcanic arc. Forearcs have the lowest heatflow from the interior Earth because there is no asthenosphere (convecting mantle) between the forearc lithosphere and the cold subducting plate.

The inner trench wall marks the edge of the overriding plate and the outermost forearc. The forearc consists of igneous and metamorphic crust, and this crust acts as buttress to a growing accretionary prism (sediments scraped off the downgoing plate onto the inner trench wall, depending on how much sediment is supplied to the trench). If the flux of sediments is high, material will be transferred from the subducting plate to the overriding plate. In this case an accretionary prism grows and the location of the trench migrates progressively away from the volcanic arc over the life of the convergent margin. Convergent margins with growing accretionary prisms are called accretionary convergent margins and make up nearly half of all convergent margins. If the sediment flux is low, material will be transferred from the overriding plate to the subducting plate by a process of tectonic ablation known as subduction erosion and carried down the subduction zone. Forearcs undergoing subduction erosion typically expose igneous rocks. In this case, the location of the trench will migrate towards the magmatic arc over the life of the convergent margin. Convergent margins experiencing subduction erosion are called nonaccretionary convergent margins and comprise more than half of convergent plate boundaries. This is an oversimplification, because different parts of a convergent margin can experience sediment accretion and subduction erosion over its life.

The asymmetric profile across a trench reflects fundamental differences in materials and tectonic evolution. The outer trench wall and outer swell comprise seafloor that takes a few million years to move from where subduction-related deformation begins near the outer trench swell until sinking beneath the trench. In contrast, the inner trench wall is deformed by plate interactions for the entire life of the convergent margin. The forearc is continuously subjected to subduction-related earthquakes. This protracted deformation and shaking ensures that the inner trench slope is controlled by the angle of repose of whatever material it is composed of. Because they are composed of igneous rocks instead of deformed sediments, non-accretionary trenches have steeper inner walls than accretionary trenches.

Filled Trenches

The composition of the inner trench slope and a first-order control on trench morphology is determined by sediment supply. Active accretionary prisms are common for trenches near continents where large rivers or glaciers reach the sea and supply great volumes of sediment which naturally flow to the trench. These filled trenches are confusing because in a plate tectonic sense they are indistinguishable from other convergent margins but lack the bathymetric expression of a trench. The Cascadia margin of the northwest USA is a filled trench, the result of sediments delivered by the rivers of the NW USA and SW Canada. The Lesser Antilles convergent margin shows the importance of proximity to sediment sources for trench morphology. In the south, near the mouth of the Orinoco River, there is no morphological trench and the forearc plus accretionary prism is almost 500 km (310 mi) wide. The accretionary prism is so large that it forms the islands of Barbados and Trinidad. Northward the forearc narrows, the accretionary prism disappears, and only north of 17°N the morphology of a trench is seen. In the extreme north, far away from sediment sources, the Puerto Rico Trench is over 8,600 m (28,200 ft) deep and there is no active accretionary prism. A similar relationship between proximity to rivers, forearc width, and trench morphology can be observed from east to west along the Alaskan-Aleutian convergent margin. The convergent plate boundary offshore Alaska changes along its strike from a filled trench with broad forearc in the east (near the coastal rivers of Alaska) to a deep trench with narrow forearc in the west (offshore the Aleutian islands). Another example is the Makran convergent margin offshore Pakistan and Iran, which is a trench filled by sediments from the Tigris-Euphrates and Indus rivers. Thick accumulations of turbidites along a trench can be supplied by down-axis transport of sediments that enter the trench 1,000–2,000 km (620–1,240 mi) away, as is found for the Peru–Chile Trench south of Valparaíso and for the Aleutian Trench. Convergence rate can also be important for controlling trench depth, especially for trenches near continents, because slow convergence causes the capacity of the convergent margin to dispose of sediment to be exceeded.

There an evolution in trench morphology can be expected as oceans close and continents converge. While the ocean is wide, the trench may be far away from continental sources of sediment and so may be deep. As the continents approach each other, the trench may become filled with continental sediments and become shallower. A simple way to approximate when the transition from subduction to collision has occurred is when the plate boundary previously marked by a trench is filled enough to rise above sealevel.

Accretionary Prisms and Sediment Transport

Accretionary prisms grow by frontal accretion, whereby sediments are scraped off, bulldozer-fashion, near the trench, or by underplating of subducted sediments and perhaps oceanic crust along the shallow parts of the subduction decollement. Frontal accretion over the life of a convergent margin results in younger sediments defining the outermost part of the accretionary prism and the oldest sediments defining the innermost portion. Older (inner) parts of the accretionary prism are much more lithified and have steeper structures than the younger (outer) parts. Underplating is difficult to detect in modern subduction zones but may be recorded in ancient accretionary prisms such as the Franciscan Group of California in the form of tectonic mélanges and duplex structures. Different modes of accretion are reflected in morphology of the inner slope of the trench, which generally shows three morphological provinces. The lower slope comprises imbricate thrust slices that form ridges. The mid slope may comprise a bench or terraces. The upper slope is smoother but may be cut by submarine canyons. Because accretionary convergent margins have high relief, are continuously deformed, and accommodate a large flux of sediments, they are vigorous systems of sediment dispersal and accumulation. Sediment transport is controlled by submarine landslides, debris flows, turbidity currents, and contourites. Submarine canyons transport sediment from beaches and rivers down the upper slope. These canyons form by channelized turbidites and generally lose definition with depth because continuous faulting disrupts the submarine channels. Sediments move down the inner trench wall via channels and a series of fault-controlled basins. The trench itself serves as an axis of sediment transport. If enough sediment moves to the trench, it may be completely filled so that turbidity currents are able to carry sediments well beyond the trench and may even surmount the outer swell. Sediments from the rivers of SW Canada and NW USA spill over where the Cascadia trench would be and cross the Juan de Fuca plate to reach the spreading ridge several hundred kilometres to the west.

The slope of the inner trench slope of an accretionary convergent margin reflects continuous adjustments to the thickness and width of the accretionary prism. The prism maintains a 'critical taper', established in conformance with Mohr–Coulomb theory for the pertinent materials. A package of sediments scraped off the downgoing lithospheric plate will deform until it and the accretionary prism that it has been added to attain a critical taper (constant slope) geometry. Once critical taper is attained, the wedge slides stably along its basal decollement. Strain rate and hydrologic properties strongly influence the strength of the accretionary prism and thus the angle of critical taper. Fluid pore pressures modify rock strength and are important controls of critical taper angle. Low permeability and rapid convergence may result in pore pressures that exceed lithostatic pressure and a relatively weak accretionary prism with a shallowly tapered geometry, whereas high permeability and slow convergence result in lower pore pressure, stronger prisms, and steeper geometry.

The Hellenic Trench of the Hellenic arc system is unusual because this convergent margin subducts evaporites. The slope of the surface of the southern flank of the Mediterranean Ridge (its accretionary prism) is low, about 1°, which indicates very low shear stress on the decollement at the base of the wedge. Evaporites influence the critical taper of the accretionary complex, as their mechanical properties differ from those of siliciclastic sediments, and because of their effect upon fluid flow and fluid pressure, which control effective stress. In the 1970s, the linear deeps of the Hellenic trench south of Crete were interpreted to be similar to trenches at other subduction zones, but with the realization that the Mediterranean Ridge is an accretionary complex, it became apparent that the Hellenic trench is actually a starved forearc basin, and that the plate boundary lies south of the Mediterranean Ridge.

Water and Biosphere

The volume of water escaping from within and beneath the forearc results in some of Earth's most dynamic and complex interactions between aqueous fluids and rocks. Most of this water is trapped in pores and fractures in the upper lithosphere and sediments of the subducting plate. The average forearc is underrun by a solid volume of oceanic sediment that is 400 m (1,300 ft) thick. This sediment enters the trench with 50-60% porosity. These sediments are progressively squeezed as they are subducted, reducing void space and forcing fluids out along the decollement and up into the overlying forearc, which may or may not have an accretionary prism. Sediments accreted to the forearc are another source of fluids. Water is also bound in hydrous minerals, especially clays and opal. Increasing pressure and temperature experienced by subducted materials converts the hydrous minerals to denser phases that contain progressively less structurally bound water. Water released by dehydration accompanying phase transitions is another source of fluids introduced to the base of the overriding plate. These fluids may travel through the accretionary prism diffusely, via interconnected pore spaces in sediments, or may follow discrete channels along faults. Sites of venting may take the form of mud volcanoes or seeps and are often associated with chemosynthetic communities. Fluids escaping from the shallowest parts of a subduction zone may also escape along the plate boundary but have rarely been observed draining along the trench axis. All of these fluids are dominated by water but also contain dissolved ions and organic molecules, especially methane. Methane is often sequestered in an ice-like form (methane clathrate, also called gas hydrate) in the forearc. These are a potential energy source and can rapidly break down. Destabilization of gas hydrates has contributed to global warming in the past and will likely do so in the future.

Chemosynthetic communities thrive where cold fluids seep out of the forearc. Cold seep communities have been discovered in inner trench slopes down to depths of 7000 m in the western Pacific, especially around Japan, in the Eastern Pacific along North, Central and South America coasts from the Aleutian to the Peru–Chile trenches, on the Barbados prism, in the Mediterranean, and in the Indian Ocean along the Makran and Sunda convergent margins. These communities receive much less attention than the chemosynthetic communities associated with hydrothermal vents. Chemosynthetic communities are located in a variety of geological settings: above over-pressured sediments in accretionary prisms where fluids are expelled through mud volcanoes or ridges (Barbados, Nankai and Cascadia); along active erosive margins with faults; and along escarpments caused by debris slides (Japan trench, Peruvian margin). Surface seeps may be linked to massive hydrate deposits and destabilization (e.g. Cascadia margin). High concentrations of methane and sulfide in the fluids escaping from the seafloor are the principal energy sources for chemosynthesis.

Empty Trenches and Subduction Erosion

Trenches distant from an influx of continental sediments lack an accretionary prism, and the inner slope of such trenches is commonly composed of igneous or metamorphic rocks. Non-accretionary convergent margins are characteristic of (but not limited to) primitive arc systems. Primitive arc systems are those built on oceanic lithosphere, such as the Izu-Bonin-Mariana, Tonga-Kermadec, and Scotia (South Sandwich) arc systems. The inner trench slope of these convergent margins exposes the crust of the forearc, including basalt, gabbro, and serpentinized mantle peridotite. These exposures allow easy access to study the lower oceanic crust and upper mantle in place and provide a unique opportunity to study the magmatic products associated with the initiation of subduction zones. Most ophiolites probably originate in a forearc environment during the initiation of subduction, and this setting favors ophiolite emplacement during collision with blocks of thickened crust. Not all non-accretionary convergent margins are associated with primitive arcs. Trenches adjacent to continents where there is little influx of sediments carried by rivers, such as the central part of the Peru–Chile Trench, may also lack an accretionary prism.

Igneous basement of a nonaccretionary forearc may be continuously exposed by subduction erosion. This transfers material from the forearc to the subducting plate and can be accomplished by frontal erosion or basal erosion. Frontal erosion is most active in the wake of seamounts being subducted beneath the forearc. Subduction of large edifices (seamount tunneling) oversteepens the forearc, causing mass failures that carry debris towards and ultimately into the trench. This debris may be deposited in graben of the downgoing plate and subducted with it. In contrast, structures resulting from subduction erosion of the base of the forearc are difficult to recognize from seismic reflection profiles, so the possibility of basal erosion is difficult to confirm. Subduction erosion may also diminish a once-robust accretionary prism if the flux of sediments to the trench diminishes.

Nonaccretionary forearcs may also be the site of serpentine mud volcanoes. These form where fluids released from the downgoing plate percolate upwards and interact with cold mantle lithosphere of the forearc. Mantle peridotite is hydrated into serpentinite, which is much less dense than peridotite and so will rise diapirically when there is an opportunity to do so. Some nonaccretionary forearcs are subjected to strong extensional stresses, for example the Marianas, and this allows buoyant serpentinite to rise to the seafloor where they form serpentinite mud volcanoes. Chemosynthetic communities are also found on non-accretionary margins such as the Marianas, where they thrive on vents associated with serpentinite mud volcanoes.

Factors Affecting Trench Depth

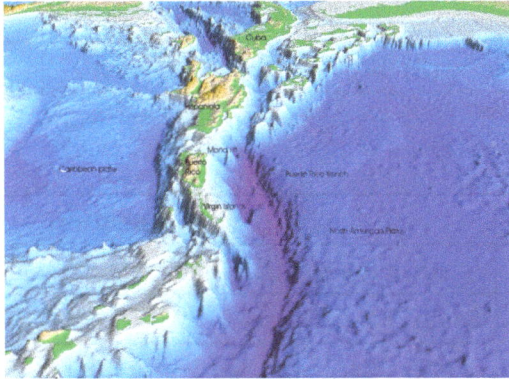

The Puerto Rico Trench

There are several factors that control the depth of trenches. The most important control is the supply of sediment, which fills the trench so that there is no bathymetric expression. It is therefore not surprising that the deepest trenches (deeper than 8,000 m (26,000 ft)) are all nonaccretionary. In contrast, all trenches with growing accretionary prisms are shallower than 8,000 m (26,000 ft). A second order control on trench depth is the age of the lithosphere at the time of subduction. Because oceanic lithosphere cools and thickens as it ages, it subsides. The older the seafloor, the deeper it lies and this determines a minimum depth from which seafloor begins its descent. This obvious correlation can be removed by looking at the relative depth, the difference between regional seafloor depth and maximum trench depth. Relative depth may be controlled by the age of the lithosphere at the trench, the convergence rate, and the dip of the subducted slab at intermediate depths. Finally, narrow slabs can sink and roll back more rapidly than broad plates, because it is easier for underlying asthenosphere to flow around the edges of the sinking plate. Such slabs may have steep dips at relatively shallow depths and so may be associated with unusually deep trenches, such as the Challenger Deep.

Deepest Oceanic Trenches

Trench	Ocean	Maximum Depth
Mariana Trench	Pacific Ocean	11,034 m (36,201 ft)
Tonga Trench	Pacific Ocean	10,882 m (35,702 ft)
Philippine Trench	Pacific Ocean	10,545 m (34,596 ft)
Kuril–Kamchatka Trench	Pacific Ocean	10,542 m (34,587 ft)
Kermadec Trench	Pacific Ocean	10,047 m (32,963 ft)
Izu-Bonin Trench (Izu-Ogasawara Trench)	Pacific Ocean	9,810 m (32,190 ft)
Japan Trench	Pacific Ocean	9,504 m (31,181 ft)

Puerto Rico Trench	Atlantic Ocean	8,800 m (28,900 ft)
South Sandwich Trench	Atlantic Ocean	8,428 m (27,651 ft)
Peru–Chile Trench or Atacama Trench	Pacific Ocean	8,065 m (26,460 ft)

Notable Oceanic Trenches

Trench	Location
Aleutian Trench	South of the Aleutian Islands, west of Alaska
Bougainville Trench	South of New Guinea
Cayman Trench	Western Caribbean Sea
Cedros Trench (inactive)	Pacific coast of Baja California
Hikurangi Trench	East of New Zealand
Izu-Ogasawara Trench	Near Izu and Bonin islands
Japan Trench	Northeast Japan
Kermadec Trench *	Northeast of New Zealand
Kuril–Kamchatka Trench *	Near Kuril islands
Manila Trench	West of Luzon, Philippines
Mariana Trench *	Western Pacific ocean; east of Mariana Islands
Middle America Trench	Eastern Pacific Ocean; off coast of Mexico, Guatemala, El Salvador, Nicaragua, Costa Rica
New Hebrides Trench	West of Vanuatu (New Hebrides Islands).
Peru–Chile Trench	Eastern Pacific ocean; off coast of Peru & Chile
Philippine Trench *	East of the Philippines
Puerto Rico Trench	Boundary of Caribbean Sea and Atlantic ocean
Puysegur trench	Southwest of New Zealand
Ryukyu Trench	Eastern edge of Japan's Ryukyu Islands
South Sandwich Trench	East of the South Sandwich Islands
Sunda Trench	Curves from south of Java to west of Sumatra and the Andaman and Nicobar Islands
Tonga Trench *	Near Tonga
Yap Trench	Western Pacific ocean; between Palau Islands and Mariana Trench

(*) The 5 deepest trenches in the world

Ancient Oceanic Trenches

Trench	Location
Intermontane Trench	Western North America; between Intermontane Islands and North America
Insular Trench	Western North America; between Insular Islands and Intermontane Islands
Farallon Trench	Western North America
Tethyan Trench	South of Turkey, Iran, Tibet and Southeast Asia

Passive Margin

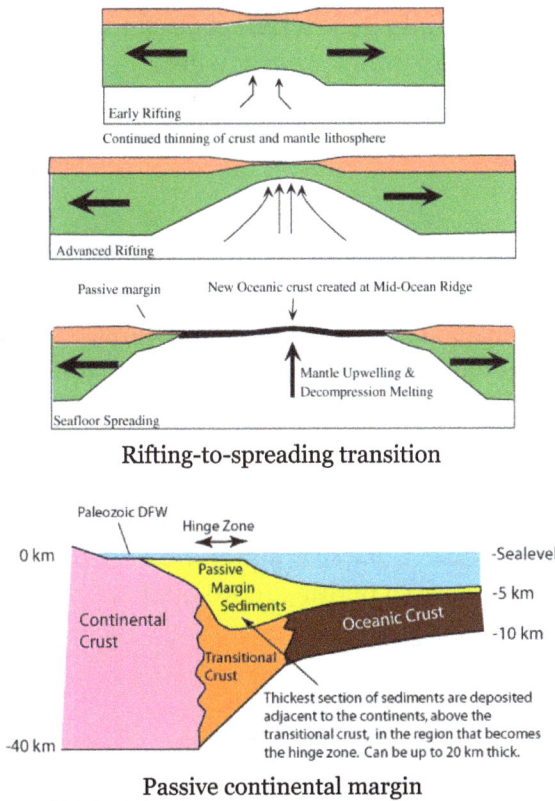

Rifting-to-spreading transition

Passive continental margin

A passive margin is the transition between oceanic and continental lithosphere that is not an active plate margin. A passive margin forms by sedimentation above an ancient rift, now marked by transitional lithosphere. Continental rifting creates new ocean basins. Eventually the continental rift forms a mid-ocean ridge and the locus of extension moves away from the continent-ocean boundary. The transition between the continental and oceanic lithosphere that was originally created by rifting is known as a passive margin.

Global Distribution

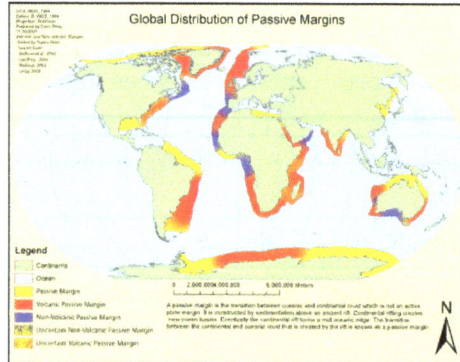

Map showing the distribution of Earth's passive margins (yellow swaths).

Passive margins are found at every ocean and continent boundary that is not marked by a strike-slip fault or a subduction zone. Passive margins define the region around the Atlantic Ocean, Arctic Ocean, and western Indian Ocean, and define the entire coasts of Africa, Greenland, India and Australia. They are also found on the east coast of North America and South America, in western Europe and most of Antarctica. East Asia also contains some passive margins.

Key Components

Active vs. Passive Margins

This refers to whether a crustal boundary between oceanic lithosphere and continental lithosphere is a plate boundary or not. Active margins are found on the edge of a continent where subduction occurs. These are often marked by uplift and volcanic mountain belts on the continental plate. Less often there is a strike-slip fault, as defines the southern coastline of W. Africa. Most of the eastern Indian Ocean and nearly all of the Pacific Ocean margin are examples of active margins. While a weld between oceanic and continental lithosphere is called a passive margin, it is not an inactive margin. Active subsidence, sedimentation, growth faulting, pore fluid formation and migration are all active processes on passive margins. Passive margins are only passive in that they are not active plate boundaries.

Morphology

Passive margins consist of both onshore coastal plain and offshore continental shelf-slope-rise triads. Coastal plains are often dominated by fluvial processes, while the continental shelf is dominated by deltaic and longshore current processes. The great rivers (Amazon. Orinoco, Congo, Nile, Ganges, Yellow, Yangtze, and Mackenzie rivers) drain across passive margins. Extensive estuaries are common on mature passive margins. Although there are many kinds of passive margins, the morphologies of most passive margins are remarkably similar. Typically they consist of a continental shelf, continental slope, continental rise, and abyssal plain. The morphological expression of these features are largely defined by the underlying transitional crust and the sedimentation above it. Passive margins defined by a large fluvial sediment budget and those dominated by coral and other biogenous processes generally have a similar morphology. In addition, the shelf break seems to mark the maximum Neogene lowstand, defined by the glacial maxima. The outer continental shelf and slope may be cut by great submarine canyons, which mark the offshore continuation of rivers.

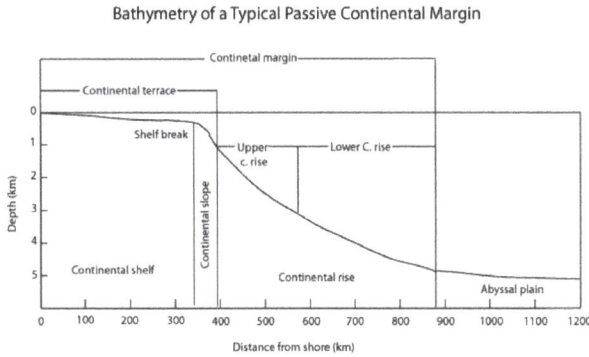

Bathymetric profile across a typical passive margin. Note that vertical scale is greatly exaggerated relative to the horizontal scale.

At high latitudes and during glaciations, the nearshore morphology of passive margins may reflect glacial processes, such as the fjords of Norway and Greenland.

Cross-section

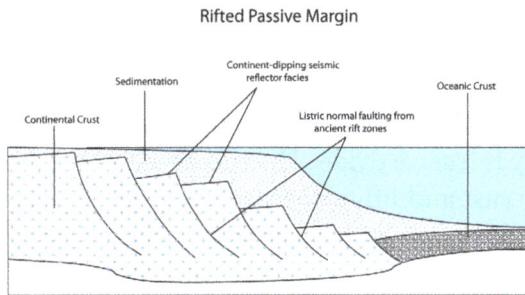

Transitional crust composed of stretched and faulted continental crust. Note: vertical scale is greatly exaggerated relative to horizontal scale.

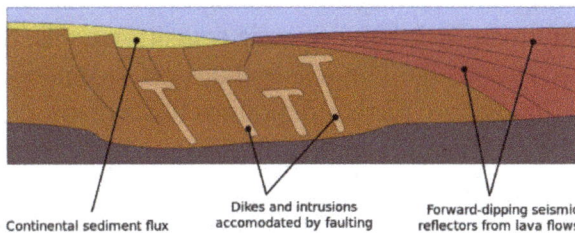

Cross-section through transitional crust of a passive margin. Transitional crust as a largely volcanic construct. Note: vertical scale is greatly exaggerated relative to horizontal scale.

The main features of passive margins lie underneath the external characters. Beneath passive margins the transition between the continental and oceanic crust is a broad transition known as transitional crust. The subsided continental crust is marked by normal faults that dip seaward. The faulted crust transitions into oceanic crust and may be deeply buried due to thermal subsidence and the mass of sediment that collects above it. The lithosphere beneath passive margins is known as transitional lithosphere. The lithosphere thins seaward as it transitions seaward to oceanic crust. Different kinds of transitional crust form, depending on how fast rifting occurs and how hot the underlying mantle was at the time of rifting. Volcanic passive margins represent one endmember transitional crust type, the other endmember (amagmatic) type is the rifted passive margin. Volcanic passive margins they also are marked by numerous dykes and igneous intrusions within the subsided continental crust. There are typically a lot of dykes formed perpendicular to the seaward-dipping lava flows and sills. Igneous intrusions within the crust cause lava flows along the top the subsided continental crust and form seaward-dipping reflectors.

Subsidence Mechanisms

Passive margins are characterized by thick accumulations of sediments. Space for these sediments is called accommodation and is due to subsidence of especially the transitional crust. Subsidence is ultimately caused by gravitational equilibrium that is established between the crustal tracts, known as isostasy. Isostasy controls the uplift of the rift flank and the subsequent subsidence of the evolving passive margin and is mostly reflected by changes in heat flow. Heat flow at passive margins changes significantly over its lifespan, high at the beginning and decreasing with age. In the initial stage, the continental crust and lithosphere is stretched and thinned due to plate movement (plate tectonics) and associated igneous activity. The very thin lithosphere beneath the rift allows the upwelling mantle to melt by decompression. Lithospheric thinning also allows the asthenosphere to rise closer to the surface, heating the overlying lithosphere by conduction and advection of heat by intrusive dykes. Heating reduces the density of the lithosphere and elevates the lower crust and lithosphere. In addition, mantle plumes may heat the lithosphere and cause prodigious igneous activity. Once a mid-oceanic ridge forms and seafoor spreading begins, the original site of rifting is separated into conjugate passive margins (for example, the eastern US and NW African margins were parts of the same rift in early Mesozoic time and are now conjugate margins) and migrates away from the zone of mantle upwelling and heating and cooling begins. The mantle lithosphere below the thinned and faulted continental oceanic transition cools, thickens, increases in density and thus begins to subside. The accumulation of sediments above the subsiding transitional crust and lithosphere further depresses the transitional crust.

Classification of Passive Margins

There are four different perspectives needed to classify passive margins:

1. map-view formation geometry (rifted, sheared, and transtensional),

2. nature of transitional crust (volcanic and non-volcanic),

3. whether the transitional crust represents a continuous change from normal continental to normal oceanic crust or this includes isolated rifts and stranded continental blocks (simple and complex), and

4. sedimentation (carbonate-dominated, clastic-dominated, or sediment starved).

The first describes the relationship between rift orientation and plate motion, the second describes the nature of transitional crust, and the third describes post-rift sedimentation. All three perspectives need to be considered in describing a passive margin. In fact, passive margins are extremely long, and vary along their length in rift geometry, nature of transitional crust, and sediment supply; it is more appropriate to subdivide individual passive margins into segments on this basis and apply the threefold classification to each segment.

Geometry of Passive Margins

Rifted Margin

This is the typical way that passive margins form, as separated continental tracts move perpendicular to the coastline. This is how the Central Atlantic opened, beginning in Jurassic time. Faulting tends to be listric: normal faults that flatten with depth.

Sheared Margin

Sheared margins form where continental breakup was associated with strike-slip faulting. A good example of this type of margin is found on the south-facing coast of west Africa. Sheared margins are highly complex and tend to be rather narrow. They also differ from rifted passive margins in structural style and thermal evolution during continental breakup. As the seafloor spreading axis moves along the margin, thermal uplift produces a ridge. This ridge traps sediments, thus allowing for thick sequences to accumulate. These types of passive margins are less volcanic.

Transtensional Margin

This type of passive margin develops where rifting is oblique to the coastline, as is now occurring in the Gulf of California.

Nature of Transitional Crust

Transitional crust, separating true oceanic and continental crusts, is the foundation of any passive margin. This forms during the rifting stage and consists of two endmembers: Volcanic and Non-Volcanic. This classification scheme only applies to rifted and transtensional margin; transitional crust of sheared margins is very poorly known.

Non-volcanic Rifted Margin

Non-volcanic margins are formed when extension is accompanied by little mantle melting and volcanism. Non-volcanic transitional crust consists of stretched and thinned continental crust. Non-volcanic margins are typically characterized by continentward-dipping seismic reflectors (rotated crustal blocks and associated sediments) and low P-wave velocities (<7.0 km/s) in the lower part of the transitional crust.

Volcanic Rifted Margin

Volcanic margins form part of large igneous provinces, which are characterised by massive emplacements of mafic extrusives and intrusive rocks over very short time periods. Volcanic margins form when rifting is accompanied by significant mantle melting, with volcanism occurring before and/or during continental breakup. The transitional crust of volcanic margins is composed of basaltic igneous rocks, including lava flows, sills, dykes, and gabbro.

Volcanic margins are usually distinguished from non-volcanic (or magma-poor) margins (e.g. the Iberian margin, Newfoundland margin) which do not contain large amounts of extrusive and/or intrusive rocks and may exhibit crustal features such as unroofed, serpentinized mantle . Volcanic margins are known to differ from magma-poor margins in a number of ways:

- a transitional crust composed of basaltic igneous rocks, including lava flows, sills, dykes, and gabbros.

- a huge volume of basalt flows, typically expressed as seaward-dipping reflector sequences (SDRS) rotated during the early stages of crustal accretion (breakup stage),

- The presence of numerous sill/dyke and vent complexes intruding into the adjacent basin,

- the lack of significant passive-margin subsidence during and after breakup, and

- the presence of a lower crust with anomalously high seismic P-wave velocities (V_p=7.1-7.8 km/s) – referred to as lower crustal bodies (LCBs) in the geologic literature.

The high velocities (V_p > 7 km) and large thicknesses of the LCBs are evidence that supports the case for plume-fed accretion (mafic thickening) underplating the crust during continental breakup. LCBs are located along the continent-ocean transition but can sometimes extend beneath the continental part of the rifted margin (as observed in the mid-Norwegian margin for example). In the continental domain, there are still open discussion on their real nature, chronology, geodynamic and petroleum implications.

Examples of volcanic margins:

- The Yemen margin

- The East Australian margin

- The West Indian margin

- The Hatton-Rockal margin

- The U.S East Coast

- The mid-Norwegian margin

- The Brazilian margins

- The Namibian margin

- The East Greenland margin

- The West Greenland margin

Examples of non-volcanic margins:

- The Newfoundland Margin

- The Iberian Margin

- The Margins of the Labrador Sea (Labrador and Southwest Greenland)

Heterogeneity of Transitional Crust

Simple Transitional Crust

Passive margins of this type show a simple progression through the transitional crust, from normal continental to normal oceanic crusts. The passive margin offshore Texas is a good example.

Complex Transitional Crust

This type of transitional crust is characterized by abandoned rifts and continental blocks, such as the Blake Plateau, Grand Banks, or Bahama Islands offshore eastern Florida.

Sedimentation

A fourth way to classify passive margins is according to the nature of sedimentation of the mature passive margin. Sedimentation continues throughout the life of a passive margin. Sedimentation changes rapidly and progressively during the initial stages of passive margin formation because rifting begins on land, becoming marine as the rift opens and a true passive margin is established. Consequently, the sedimentation history of a passive margin begins with fluvial, lacustrine, or other subaerial deposits, evolving with time depending on how the rifting occurred and how, when, and by what type of sediment it varies.

Constructional

Constructional margins are the "classic" mode of passive margin sedimentation. Normal sedimentation results from the transport and deposition of sand, silt, and clay by rivers via deltas and redistribution of these sediments by longshore currents. The nature of sediments can change remarkably along a passive margin, due to interactions between carbonate sediment production, clastic input from rivers, and alongshore transport. Where clastic sediment inputs are small, biogenic sedimentation can dominate especially nearshore sedimentation. The Gulf of Mexico passive margin along the southern United States is an excellent example of this, with muddy and sandy coastal environments down current (west) from the Mississippi River Delta and beaches of carbonate sand to the east. The thick layers of sediment gradually thin with increasing distance offshore, depending on subsidence of the passive margin and the efficacy of offshore transport mechanisms such as turbidity currents and submarine channels.

Development of the shelf edge and its migration through time is critical to the development of a passive margin. The location of the shelf edge break reflects complex interaction between sedimentation, sealevel, and the presence of sediment dams. Coral reefs serve as bulwarks that allow sediment to accumulate between them and the shore, cutting off sediment supply to deeper water. Another type of sediment dam results from the presence of salt domes, as are common along the Texas and Louisiana passive margin.

Starved

Sediment-starved margins produce narrow continental shelves and passive margins. This is especially common in arid regions, where there is little transport of sediment by rivers or redistribution by longshore currents. The Red Sea is a good example of a sediment-starved passive margin.

Formation

There are three main stages in the formation of passive margins:

1. In the first stage a continental rift is established due to stretching and thinning of the crust and lithosphere by plate movement. This is the beginning of the continental crust subsidence. Drainage is generally away from the rift at this stage.

2. The second stage leads to the formation of an oceanic basin, similar to the modern Red Sea. The subsiding continental crust undergoes normal faulting

as transitional marine conditions are established. Areas with restricted sea water circulation coupled with arid climate create evaporite deposits. Crust and lithosphere stretching and thinning are still taking place in this stage. Volcanic passive margins also have igneous intrusions and dykes during this stage.

3. The last stage in formation happens only when crustal stretching ceases and the transitional crust and lithosphere subsides as a result of cooling and thickening (thermal subsidence). Drainage starts flowing towards the passive margin causing sediment to accumulate over it.

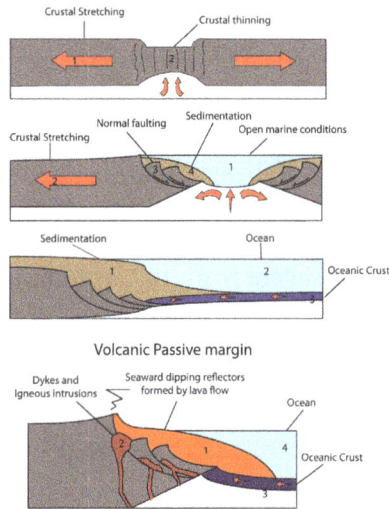

Economic Significance

Passive margins are important exploration targets for petroleum. Mann et al. (2001) classified 592 giant oil fields into six basin and tectonic-setting categories, and noted that continental passive margins account for 31% of giants. Continental rifts (which are likely to evolve into passive margins with time) contain another 30% of the world's giant oil fields. Basins associated with collision zones and subduction zones are where most of the remaining giant oil fields are found.

Passive margins are petroleum storehouses because these are associated with favorable conditions for accumulation and maturation of organic matter. Early continental rifting conditions led to the development of anoxic basins, large sediment and organic flux, and the preservation of organic matter that led to oil and gas deposits. Crude oil will form from these deposits. These are the localities in which petroleum resources are most profitable and productive. Productive fields are found in passive margins around the globe, including the Gulf of Mexico, western Scandinavia, and Western Australia.

Law of the Sea

International discussions about who controls the resources of passive margins are the focus of Law of the Sea negotiations. Continental shelves are important parts of national exclusive economic zones, important for seafloor mineral deposits (including oil and gas) and fisheries.

Volcanic Passive Margin

Volcanic passive margins (VPM) and non-volcanic passive margins are the two forms of transitional crust that lie beneath passive continental margins that occur on Earth as the result of the formation of ocean basins via continental rifting. Initiation of igneous processes associated with volcanic passive margins occurs before and/or during the rifting process depending on the cause of rifting. There are two accepted models for VPM formation: hotspots/mantle plumes and slab pull. Both result in large, quick lava flows over a relatively short period of geologic time (i.e. a couple of million years). VPM's progress further as cooling and subsidence begins as the margins give way to formation of normal oceanic crust from the widening rifts.

Characteristics

Despite the differences in origin and formation, most VPMs share the same characteristics:

- 4 to 7 km thick basaltic and (frequently) silicic subaerial flows; dike swarms and sills running parallel to continent-facing normal faults.

- 10 to 15 km thick bodies in the lower crust (HVLC) show high seismic P-Wave velocities, between 7.1 and 7.8 km/s which lie under the transitional crust (crust between continental crust and oceanic crust).

- Seaward Dipping Reflector (SDR) series: Inner SDR's overlay transitional continental crust. They are composed of varying mixtures of subaerial volcanic flows, volcaniclastic and non-volcanic sediments which range from 50–150 km wide and are 5–10 km thick. Outer SDR's overlay transitional oceanic crust are composed of submarine basaltic flows which range from 3 to 9 km thick.

Development

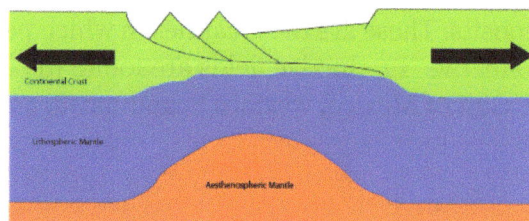

Not to scale   Extensional stress leads to asthenospheric upwelling and Listric Faulting.

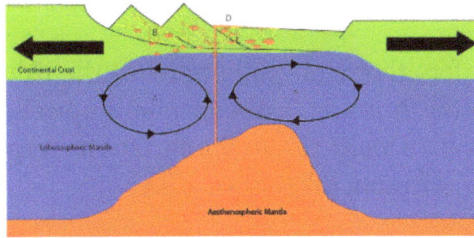

Not to scale   Asthenospheric upwelling, listric faulting, and crustal thinning continue. Mantle convection (A) further weakens lithosphere and leads to the formation of dikes and sills (B). Dikes and sills feed magma chambers in the lower and upper crust (C). Lava erupts as basaltic sheet flows (D).

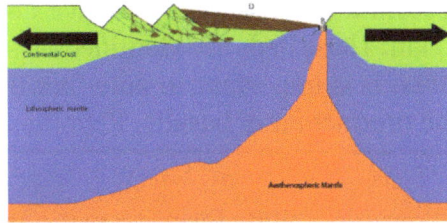

Not to scale   Thinning crust is strained to the point of breaking, forming a mid-ocean ridge (A). Mantle material upwells to fill the gap at the mid-ocean ridge (B) and cools to form oceanic crust (C). Volcanic sheet flows atop transitional oceanic crust form outer seaward dipping reflectors (D). Convecting mantle material along base of transitional crust cools to form HVLC (E).

Magma Plume VPM

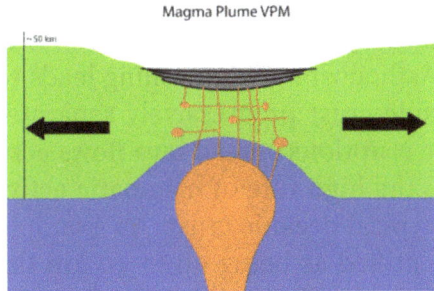

Extension thins the crust. Magma reaches the surface through radiating sills and dikes, forming basalt flows, as well as deep and shallow magma chambers below the surface. The crust gradually sink due to thermal subsidence, and originally horizontal basalt flows are rotated tosees become seaward dipping reflectors.

Rift Initiation

Active Rifting

The active rift model sees rupture driven by hotspot or mantle plume activity. Upwellings of hot mantle, known as mantle plumes, originate deep in Earth and rise to heat and thin the lithosphere. Heated lithosphere thins, weakens, rises, and finally rifts, Enhanced melting following continental breakup is very important in VPMs, creating thicker than normal oceanic crust of 20 to 40 km thick. Other melts caused by convection related upwelling form reservoirs of magma from which dike swarms and sills eventually radiate to the surface, creating the characteristic seaward dipping lava flows. This model is controversial.

Passive Rifting

The passive rift model infers that slab pull stretches the lithosphere and thins it. To compensate for lithospheric thinning, asthenosphere upwells, melts due to adiabatic decompression, and derivative melts rise to the surface to erupt. Melts push up through faults towards the surface, forming dikes and sills.

Development of Transitional Crust

Continued extension leads to accelerated igneous activity, including repeated eruptions. Repeated eruptions form a thick sequence of lava beds that can reach a combined thickness of up to 20 km. These beds are identified on seismic refraction sections as seaward dipping reflectors. It is important to note that the early phase of volcanic activity is not limited to the production of basalts. Rhyolite and other felsic rocks can also be found in these zones.

Continued extension with volcanic activity forms transitional crust, welding ruptured continent to nascent ocean floor. Volcanic beds cover the transition from thinned continental crust to oceanic crust. Also occurring during this phase is the formation of high velocity seismic zones under the thinned continental crust and the transition crust. These zones are identified by typical seismic velocities between 7.2-7.7 km/s and are usually interpreted as layers of mafic to ultramafic rocks that have underplated the transitional crust. Asthenospheric upwelling leads to the formation of a mid-ocean ridge and new oceanic crust progressively separates the once-conjoined rift halves. Continued volcanic eruptions spread lava flows across transitional crust and onto oceanic crust. Due to the high rate of magmatic activity the new oceanic crust forms much thicker than typical oceanic crust. An example of this is Iceland where oceanic crust has been identified as being up to 40 km thick. Some have theorized that the copious amounts of volcanic material also lead to the formation of oceanic plateaus at this time.

Post-rift

The final and longest phase is the continued thermal subsidence of the transitional crust and the accumulation of sediments. Continued seafloor spreading leads to the formation of oceanic crust of normal thickness. Over time this production of normal oceanic crust and sea floor spreading leads to the formation of an ocean. This phase is of the most interest to the oil industry and sedimentary geologists.

Distribution and Examples

The distribution of known volcanic margins is shown on the graphic to the right. Many of the margins have not been thoroughly investigated and more passive margins are identified as volcanic from time to time.

Volcanic passive margins:

- South Atlantic

- Western Australia

- Southwest India

- West Greenland

- East Greenland

- Northern Labrador Sea

- South of Arabia

- Norwegian Margin

- US Atlantic Margin

VPM Example: The US Atlantic Margin

The US Atlantic passive margin extends from Florida to southern Nova Scotia. This VPM was a result of the breakup of the supercontinent, Pangea, in which North America separated from northwestern Africa and Iberia to form the North Atlantic Ocean. This margin has a typical history of tectonic events that are representative of volcanic passive margins with rifting and passive margin formation occurreing 225-165 million years ago. Like other VPMs the US East Coast Margin developed in two stages; 1) rifting, initiated during the Middle to Late Triassic and continued into Jurassic time and 2) seafloor spreading, which began in Jurassic time and continues today. The US East Coast includes several components which are characteristic of VPM's including; seaward-dipping reflectors, flood basalts, dikes, and sills.

Non-volcanic Passive Margins

Non-volcanic passive margins (NVPM) constitute one end member of the transitional crustal types that lie beneath passive continental margins; the other end member being volcanic passive margins (VPM). Transitional crust welds continental crust to oceanic crust along the lines of continental break-up. Both VPM and NVPM form during rifting, when a continent rifts to form a new ocean basin. NVPM are different from VPM because of a lack of volcanism. Instead of intrusive magmatic structures, the transitional crust is composed of stretched continental crust and exhumed upper mantle. NVPM are typically submerged and buried beneath thick sediments, so they must be studied using geophysical techniques or drilling.

NVPM have diagnostic seismic, gravity, and magnetic characteristics that can be used to distinguish them from VPM and for demarcating the transition between continental and oceanic crust.

Typical Characteristics

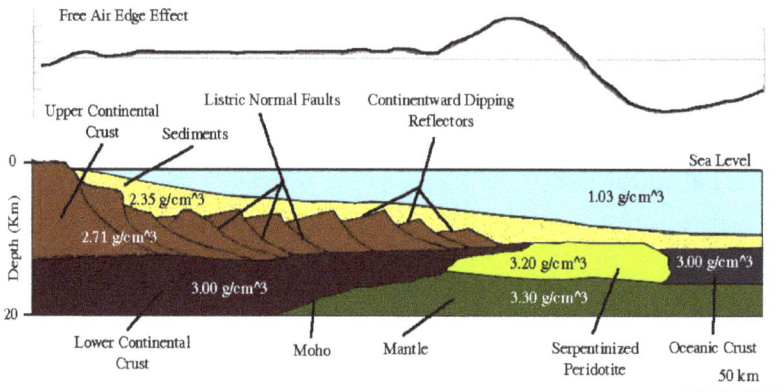

NVPM are the result of rifting when a continent breaks up to form an ocean, producing transitional crust without volcanism. Extension causes a number of events to occur. First is lithospheric thinning, which allows asthenospherc upwelling; heating further erodes the lithosphere, furthering the thinning process. The extensional forces also cause listric faults and continentward dipping reflectors that help identify NVPM and distinguish them from VPM, characterized by seaward-dipping seismic reflectors. The main difference between NVPM and VPM is that in the latter case, the mantle is hot enough to melt and produce voluminous basalts, whereas in the former case the mantle doesn't melt and there is little or no volcanism. Instead, extension simply pulls the crust away, exposing or "unroofing" the mantle, exposing serpentinized peridotite. The mantle doesn't melt because it is cold or upwells slowly, so there are no igneous rocks like there are in VPM. The basalts and granites are replaced with serpentinized peridotite, accompanied by unique serpentothemal and hydrothermal activity. Increasing density of the lithosphere as it cools and sediment accumulation causes subsidence.

Geophysical Properties

Seismic Characteristics

Seismic reflection lines across passive margins show many structural features common to both VPM and NVPM, such as faulting and crustal thinning, with the primary contra-indicator for volcanism being the presence of continent-ward dipping reflectors.

NVPM also display distinct p-wave velocity structures that differentiate them from VPM. Typical NVPM exhibit a high velocity, high gradient lower crust (6.4-

7.7 km/s) overlain by a thin, low velocity (4–5 km/s) upper crustal layer. The high velocity shallow layer is usually interpreted as the serpentinized peridotite associated with NVPM. In some cases, an extremely thick igneous underplating of a VPM will display similar P-wave velocity (7.2-7.8 km/s, but with a lower gradient). For this reason, velocity structure alone cannot be used to determine the nature of a margin.

Gravity Properties

Gravity data provides information about the subsurface density distribution. The most important gravity feature associated with any continent-ocean transition, including NVPM, is the free-air edge effect anomaly, which consists of a gravity high and a gravity low associated with the contrast between the thick continental and thin oceanic crust. There are also subsurface variations in density that cause significant variations across the continent-ocean transition. The crust, as well as the entire lithosphere, is thinned due to mechanical extension. The Moho marks a large density contrast between crust and mantle, typically at least 0.35 g/cm3. The highest amplitudes of the gravity anomaly occur seaward of the continent-ocean transition. High-density upper mantle material is elevated relative to the more landward crustal root. The oceanic crust density is then further enhanced with gabbros and basalts and additionally contributes to the regional gravity trend.

Where the thickness of the crust and lithosphere varies, equilibrium must be reached. Isostatic compensation and gravity anomalies result from balance between mass excess of the extra mantle beneath the thinned lithosphere and the overlying low-density crust. Positive gravity anomalies result from the relatively low flexural strength of the lithosphere during the beginning of rifting. As the passive margin matures, the crust and uppermost mantle become colder and stronger, so that the compensating deflection in the base of the lithosphere is broader than the actual rift. Higher flexural strength results in a broadening of the gravity anomaly with time.

Magnetic Properties

The magnetic signature of a passive continental margin is influenced by the volume of material with a high magnetic susceptibility and the depth of the material below the surface. Large amplitude magnetic anomalies are associated with high magnetic susceptibility (~0.06 emu) igneous rocks of VPM. In contrast, NVPM exhibit only small amplitude anomalies associated with the edge effect at the boundary between the exhumed mantle (~0.003 emu) at the transition zone, and the true oceanic crust basalt (~0.05 emu). This anomaly can be used to locate the boundary between transitional crust and oceanic crust. The absence of large amplitude anomalies is a very strong indication that a margin is non-volcanic.

Formation

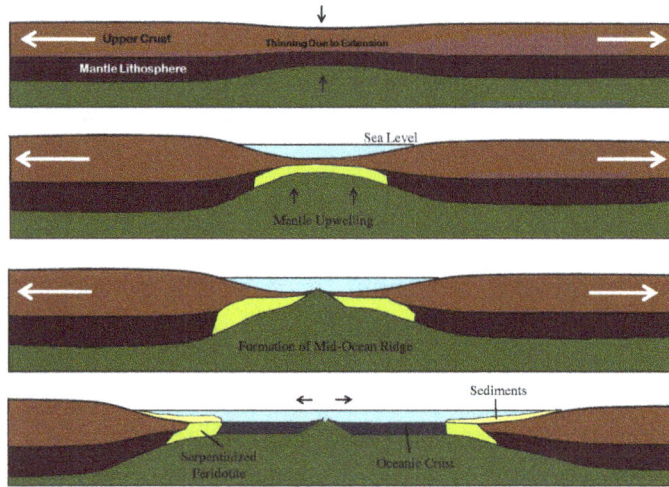

Passive Rifting

Passive rifting, unlike active rifting, occurs principally by extensional tectonic forces as opposed to magmatic forces originating from convection cells or mantle plumes. Isostatic forces allow mantle material to rise under the thinning lithosphere. Subsidence and sedimentation occur during both the initial rifting stage and the post rifting stages. Only after initial rifting does any mantle melting occur. Continued extension of the lithosphere will eventually lead to decompression melting of the mantle and the formation of a mid-ocean ridge. This process results in the creation of an ocean basin, and possibly conjugate NVPM (Geoffroy & 200).

Rifting Models

There are several models for forming NVPM. Passive rifting can follow McKenzie's pure shear model, Wernicke's simple shear model, or a composite model combining features of both, as observed at the Galicia bank NVPM.

Mckenzie Pure Shear Model

Pure shear describes "homogeneous flattening" of rocks without rotations, while maintaining a constant volume. If a cube undergoes pure shearing, the result will be a rectangular prism with sides parallel to those of the initial cube. McKenzie's model predicts symmetric structures on either side of the rift zone composed of rotated fault blocks bounded by normal faults (McKenzie 1978).

Wernicke Simple Shear Model

In contrast to pure shear, simple shear describes constant volume strain with rotations. If a cube undergoes simple shearing, the result will be a parallelogram with sides that increase in length and are no longer parallel to the sides of the original cube. The top and bottom of the cube will neither stretch nor shorten. In a simple shear model, a basin is stretched asymmetrically by a large scale detachment fault extending from the upper crust to the lower lithosphere and even asthenosphere (Wernicke 1985).

Galicia Bank

Composite Model Formation

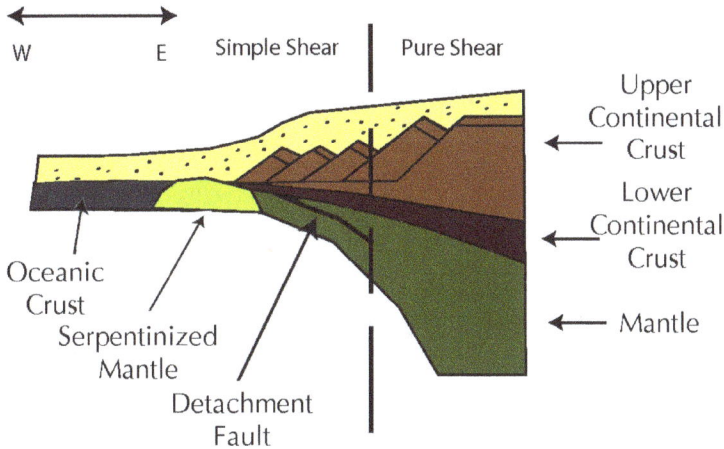

During the Late Jurassic-Early Cretaceous, tectonic extensional forces created a shallow angle east-dipping detachment fault. This fault cut from what is now the Flemish Cap margin in Nova Scotia, eastern Canada to the Galicia margin, which is located west of the Iberian Peninsula. This fault penetrated the upper portion of the continental crust and merged into the transition between brittle upper and plastic lower crust. In time, displacement along this detachment fault decreased to zero at a point under the Galicia margin. East of this detachment fault, the structure of the Galicia NVPM is entirely pure shear resulting in rotated fault blocks, normal faults, and continent-ward dipping seismic reflectors. Simple shear is only evident in the western edge of the Galicia margin and the upper crust of the Flemish Cap margin where the crust is brittle. Below this brittle crust, the ductile crust follows McKenzie's pure shear model. Mantle material composed of peridotites is serpentinized by circulating seawater after it rises close enough to the upper crust due to its low density and isostatic forces. After sufficient thinning of the lithosphere, this serpentinized material is emplaced at the continent-ocean transition. This is why the transitional crust of NVPM are made of serpentinized peridotite instead of magmatic structures seen in VPM. Since the emplacement of the peridotite, oceanic crust has been forming at the Mid-Atlantic Ridge and driving the two NVPM apart. The simple shear detachment became a deactivated detachment fault once this rifting process began the formation of new oceanic crust. This process explains the structures seen at the Galicia margin today.

References

- Hsü, Kenneth J. (1992). Challenger at Sea: A Ship That Revolutionized Earth Science. Princeton University Press. p. 57. ISBN 0-691-08735-0.

- Redfern, R.; 2001: Origins, the Evolution of Continents, Oceans and Life, University of Oklahoma Press, ISBN 1-84188-192-9, p. 26

- Spencer, Edgar W. (1977). Introduction to the Structure of the Earth (2nd ed.). Tokyo: McGraw-Hill. ISBN 0-07-085751-2.

- Bhagwat, S.B. (2009). Foundation of Geology Vol 1. Global Vision Publishing House. p. 83. ISBN 9788182202764.

- Moores, Eldridge M.; Twiss, Robert J. (1995). Tectonics. W.H. Freeman Company. pp. 16–20, 97–104. ISBN 0-7167-2437-5.

- Burke, K.; Dewey, J. F. (1973). "Plume-generated triple junctions: key indicators in applying plate tectonics to old rocks" (PDF). The Journal of Geology: 406–433. Retrieved 23 October 2016.

- Burke, K. (1976). "Development of graben associated with the initial ruptures of the Atlantic Ocean" (PDF). Tectonophysics. 36 (1-3): 93–112. Retrieved 23 October 2016.

- Pérez-Díaz & Eagles 2014, Opening of Equatorial and High-Latitude Gateways to the South Atlantic, pp. 18–19

- Harpp, Karen (2001). "Tracing a mantle plume: Isotopic and trace element variations of Galapagos seamounts" (PDF). Geochemistry, Geophysics, Geosystems. Retrieved 2011-04-06.

8

Supercontinents: An Integrated Study

A supercontinent is the collected landmass of almost all of the Earth's continents. An ancient supercontinent mentioned in the text is Gondwana. This section serves as a source to understand all the major supercontinents, and provides a better understanding on the subject matter.

Supercontinent

The Eurasian landmass would *not* be considered a supercontinent according to P.F. Hoffman (1999).

In geology, a supercontinent is the assembly of most or all of Earth's continental blocks or cratons to form a single large landmass. However, the definition of a supercontinent can be ambiguous. Many earth scientists, such as P.F. Hoffman (1999), use the term "supercontinent" to mean "a clustering of nearly all continents". This definition leaves room for interpretation when labeling a continental body and is easier to apply to Precambrian times. Using the first definition provided here, Gondwana (aka Gondwanaland) is not considered a supercontinent, because the landmasses of Baltica, Laurentia and Siberia also existed at the same time but physically separate from each other. The landmass of Pangaea is the collective name describing all of these continental masses when they were in proximity to one another. This would classify Pangaea as a supercontinent. According to the definition by Rogers and Santosh (2004), a supercontinent does not exist today. Supercontinents have assembled and dispersed multiple times in the geologic past. The positions of continents have been accurately determined back to the early Jurassic. However, beyond 200 Ma, continental positions are much less certain.

Supercontinents Throughout Geologic History

The following table displays historical supercontinents, using a general definition.

Supercontinent name	Age (Ma: millions of years ago)
Ur (Vaalbara)	~3,600–2,800
Kenorland	~2,700–2,100
Protopangea-Paleopangea	~ 2,700–2,600
Columbia (Nuna)	~1,800–1,500
Rodinia	~1,250–750
Pannotia	~600
Pangaea	~300

General Chronology

There are two contrasting models for supercontinent evolution through geological time. The first model theorizes that at least two separate supercontinents existed comprising Vaalbara (from ~3600 to 2500 Ma) and Kenorland (from ~2700 to 2450 Ma). The Neoarchean supercontinent consisted of Superia and Sclavia. These parts of Neoarchean age broke off at ~2300 and 2090 Ma and portions of them later collided to form Nuna (Northern Europe North America) (~1750 Ma). Nuna continued to develop during the Mesoproterozoic, primarily by lateral accretion of juvenile arcs, and in ~1000 Ma Nuna collided with other land masses, forming Rodinia. Between ~800 and 700 Ma Rodinia broke apart. However, before completely breaking up, some fragments of Rodinia had already come together to form Gondwana (also known as Gondwanaland) by ~530 Ma. Pangaea formed by ~300 Ma through the collision of Gondwana, Laurentia, Baltica, and Siberia.

The second model (Protopangea-Paleopangea) is based on both palaeomagnetic and geological evidence and proposes that the continental crust comprised a single supercontinent from ~2.7 Ga until break-up during the Ediacaran Period after ~0.6 Ga. The reconstruction is derived from the observation that palaeomagnetic poles converge to quasi-static positions for long intervals between ~2.7–2.2, 1.5–1.25, and 0.75–0.6 Ga with only small peripheral modifications to the reconstruction. During the intervening periods, the poles conform to a unified apparent polar wander path. Because this model shows that exceptional demands on the paleomagnetic data are satisfied by prolonged quasi-integrity, it must be regarded as superseding the first model proposing multiple diverse continents, although the first phase (Protopangea) essentially incorporates Vaalbara and Kenorland of the first model. The explanation for the prolonged duration of the Protopangea-Paleopangea supercontinent appears to be that Lid Tectonics (comparable to the tectonics operating on Mars and Venus) prevailed during Precambrian times. Plate Tectonics as seen on the contemporary Earth became dominant only during the latter part of geological times.

The Phanerozoic supercontinent Pangaea began to break up 180 Ma and is still doing so today. Because Pangaea is the most recent of Earth's supercontinents, it is the most well known and understood. Contributing to Pangaea's popularity in the classroom is the fact that its reconstruction is almost as simple as fitting the present continents bordering the Atlantic-type oceans like puzzle pieces.

Supercontinent Cycles

A supercontinent cycle is the break-up of one supercontinent and the development of another, which takes place on a global scale. Supercontinent cycles are not the same as the Wilson cycle, which is the opening and closing of an individual oceanic basin. The Wilson cycle rarely synchronizes with the timing of a supercontinent cycle. However, supercontinent cycles and Wilson cycles were both involved in the creation of Pangaea and Rodinia.

Secular trends such as carbonatites, granulites, eclogites, and greenstone belt deformation events are all possible indicators of Precambrian supercontinent cyclicity, although the Protopangea-Paleopangea solution implies that Phanerozoic style of supercontinent cycles did not operate during these times. Also there are instances where these secular trends have a weak, uneven or lack of imprint on the supercontinent cycle; secular methods for supercontinent reconstruction will produce results that have only one explanation and each explanation for a trend must fit in with the rest.

Supercontinents and Volcanism

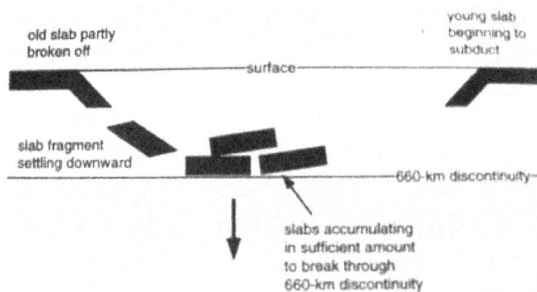

As the slab is subducted into the mantle, the more dense material will break off and sink to the lower mantle creating a discontinuity elsewhere known as a slab avalanche.

The causes of supercontinent assembly and dispersal are thought to be driven by processes in the mantle. Approximately 660 km into the mantle, a discontinuity occurs, affecting the surface crust through processes like plumes and "superplumes". When a slab of crust that is subducted is denser than the surrounding mantle, it sinks to the discontinuity. Once the slabs build up, they will sink through to the lower mantle in what is known as a "slab avalanche". This displacement at the discontinuity will cause the lower mantle to compensate and rise elsewhere. The rising mantle can form a plume or superplume.

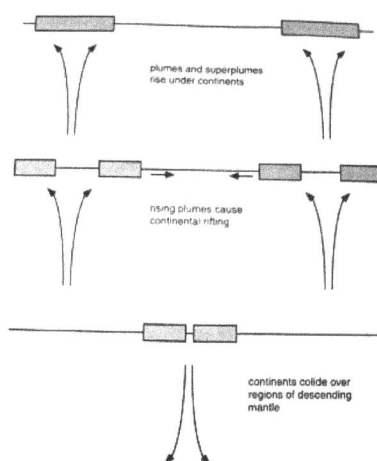

The effects of mantle plumes possibly caused by slab avalanches elsewhere in the lower mantle on the breakup and assembly of supercontinents.

Besides having compositional effects on the upper mantle by replenishing the large-ion lithophile elements, volcanism affects the plate movement. The plates will be moved towards a geoidal low perhaps where the slab avalanche occurred and pushed away from the geoidal high that can be caused by the plumes or superplumes. This causes the continents to push together to form supercontinents and was evidently the process that operated to cause the early continental crust to aggregate into Protopangea. Dispersal of supercontinents is caused by the accumulation of heat underneath the crust due to the rising of very large convection cells or plumes, and a massive heat release resulted in the final break-up of Paleopangea. Accretion occurs over geoidal lows that can be caused by avalanche slabs or the downgoing limbs of convection cells. Evidence of the accretion and dispersion of supercontinents is seen in the geological rock record.

The influence of known volcanic eruptions does not compare to that of flood basalts. The timing of flood basalts has corresponded with large-scale continental break-up. However, due to a lack of data on the time required to produce flood basalts, the climatic impact is difficult to quantify. The timing of a single lava flow is also undetermined. These are important factors on how flood basalts influenced paleoclimate.

Supercontinents and Plate Tectonics

Global paleogeography and plate interactions as far back as Pangaea are relatively well understood today. However, the evidence becomes more sparse further back in geologic history. Marine magnetic anomalies, passive margin match-ups, geologic interpretation of orogenic belts, paleomagnetism, paleobiogeography of fossils, and distribution of climatically sensitive strata are all methods to obtain evidence for continent locality and indicators of environment throughout time.

Phanerozoic (540 Ma to present) and Precambrian (4.6 Ga to 540 Ma) had primarily passive margins and detrital zircons (and orogenic granites), whereas the tenure of Pangaea contained few. Matching edges of continents are where passive margins form. The edges of these continents may rift. At this point, seafloor spreading becomes the driving force. Passive margins are therefore born during the break-up of supercontinents and die during supercontinent assembly. Pangaea's supercontinent cycle is a good example for the efficiency of using the presence, or lack of, these entities to record the development, tenure, and break-up of supercontinents. There is a sharp decrease in passive margins between 500 and 350 Ma during the timing of Pangaea's assembly. The tenure of Pangaea is marked by a low number of passive margins during 300 to 275 Ma, and its break-up is indicated accurately by an increase in passive margins.

Orogenic belts can form during the assembly of continents and supercontinents. The orogenic belts present on continental blocks are classified into three different categories and have implications of interpreting geologic bodies. Intercratonic orogenic belts are characteristic of ocean basin closure. Clear indicators of intercratonic activity contain ophiolites and other oceanic materials that are present in the suture zone. Intracratonic orogenic belts occur as thrust belts and do not contain any oceanic material. However, the absence of ophiolites is not strong evidence for intracratonic belts, because the oceanic material can be squeezed out and eroded away in an intercratonic environment. The third kind of orogenic belt is a confined orogenic belt which is the closure of small basins. The assembly of a supercontinent would have to show intercratonic orogenic belts. However, interpretation of orogenic belts can be difficult.

The collision of Gondwana and Laurasia occurred in the late Palaeozoic. By this collision, the Variscan mountain range was created, along the equator. This 6000-km-long mountain range is usually referred to in two parts: the Hercynian mountain range of the late Carboniferous makes up the eastern part, and the western part is called the Appalachians, uplifted in the early Permian. (The existence of a flat elevated plateau like the Tibetan Plateau is under much debate.) The locality of the Variscan range made it influential to both the northern and southern hemispheres. The elevation of the Appalachians would greatly influence global atmospheric circulation.

Supercontinental Climate

Continents, in particular large or supercontinents, will affect the climate of the planet drastically. In general the interaction of supercontinents and climate is similar to the interaction between present-day continents and climate, just on a different scale. Supercontinents have a larger effect on climate than do continents. The configuration and placement of the continents has a larger influence on climate. Continents modify global wind patterns, control ocean current paths and have a higher albedo than the oceans. Because continents are higher in the elevation, the temperature decreases with altitude. The wind is redirected by mountains. The albedo difference causes a shift in climate by onshore winds. "Continentality" occurs because the center of large continents are generally higher in elevations and are therefore cooler and dryer. This is seen today with Eurasia, and evidence is present in the rock record that this is true for the middle of Pangaea.

Glacial

The term glacio-epoch refers to a long episode of glaciation on Earth over millions of years. Glaciers have major implications on the climate particularly through sea level change. Changes in the position and elevation of the continents, the paleolatitude and ocean circulation affect the glacio-epochs. There is an association between the rifting and breakup of continents and supercontinents and glacio-epochs. According to the first model for Precambrian supercontinents described above the breakup of Kenorland and Rodinia were associated with the Paleoproterozoic and Neoproterozoic glacio-epochs, respectively. In contrast, the second solution described above shows that these glaciations correlated with periods of low continental velocity and it is concluded that a fall in tectonic and corresponding volcanic activity was responsible for these intervals of global frigidity. During the accumulation of supercontinents with times of regional uplift, glacio-epochs seem to be rare with little supporting evidence. However, the lack of evidence does not allow for the conclusion that glacio-epochs are not associated with collisional assembly of supercontinents. This could just represent a preservation bias.

During the late Ordovician (~465 Ma), the particular configuration of Gondwana may have allowed for glaciation and high CO_2 levels to occur at the same time. However, some geologists disagree and think that there was a temperature increase at this time. This increase may have been strongly influenced by the movement of Gondwana across the South Pole, which may have prevented lengthy snow accumulation. Although late Ordovician temperatures at the South Pole may have reached freezing, there were no ice sheets during the early Silurian (~440 Ma) through the late Mississippian (~330 Ma). Agreement can be met with the theory that continental snow can occur when the edge of a continent is near the pole. Therefore, Gondwana, although located tangent to the South Pole, may have experienced glaciation along its coast.

Precipitation

Though precipitation rates during monsoonal circulations are difficult to predict, there is evidence for a large orographic barrier within the interior of Pangaea during the late Paleozoic (~250 Ma). The possibility of the SW-NE trending Appalachian-Hercynian Mountains makes the region's monsoonal circulations potentially relatable to present day monsoonal circulations surrounding the Tibetan Plateau, which is known to positively influence the magnitude of monsoonal periods within Eurasia. It is therefore somewhat expected that lower topography in other regions of the supercontinent during the Jurassic would negatively influence precipitation variations. The breakup of supercontinents may have affected local precipitation. When any supercontinent breaks up, there will be an increase in precipitation runoff over the surface of the continental land masses, increasing silicate weathering and the consumption of CO_2.

Temperature

Even though during the Archaean solar radiation was reduced by 30 percent and the Cambrian-Precambrian boundary by six percent, the Earth has only experienced three ice ages throughout the Precambrian. It must be noted that erroneous conclusions are more likely to be made when models are limited to one climatic configuration (which is usually present day).

Cold winters in continental interiors are due to rate ratios of radiative cooling (greater) and heat transport from continental rims. To raise winter temperatures within continental interiors, the rate of heat transport must increase to become greater than the rate of radiative cooling. Through climate models, alterations in atmospheric CO_2 content and ocean heat transport are not comparatively effective.

CO_2 models suggest that values were low in the late Cenozoic and Carboniferous-Permian glaciations. Although early Paleozoic values are much larger (more than ten percent higher than that of today). This may be due to high seafloor spreading rates after the breakup of Precambrian supercontinents and the lack of land plants as a carbon sink.

During the late Permian, it is expected that seasonal Pangaean temperatures varied drastically. Subtropic summer temperatures were warmer than that of today by as much as 6–10 degrees and mid-latitudes in the winter were less than –30 degrees Celsius. These seasonal changes within the supercontinent were influenced by the large size of Pangaea. And, just like today, coastal regions experienced much less variation.

During the Jurassic, summer temperatures did not rise above zero degrees Celsius along the northern rim of Laurasia, which was the northernmost part of Pangaea (the southernmost portion of Pangaea was Gondwana). Ice-rafted dropstones sourced from Russia are indicators of this northern boundary. The Jurassic is thought to have been approximately 10 degrees Celsius warmer along 90 degrees East paleolongitude compared to the present temperature of today's central Eurasia.

Milankovitch Cycles

Many studies of the Milankovitch fluctuations during supercontinent time periods have focused on the Mid-Cretaceous. Present amplitudes of Milankovitch cycles over present day Eurasia may be mirrored in both the southern and northern hemispheres of the supercontinent Pangaea. Climate modeling shows that summer fluctuations varied 14–16 degrees Celsius on Pangaea, which is similar or slightly higher than summer temperatures of Eurasia during the Pleistocene. The largest-amplitude Milankovitch cycles are expected to have been at mid- to high-latitudes during the Triassic and Jurassic.

Proxies

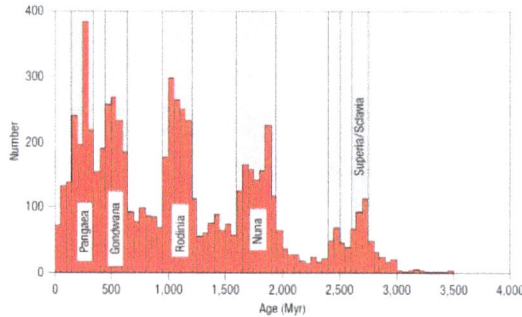

U–Pb ages of 5,246 concordant detrital zircons from 40 of Earth's major rivers

Granites and detrital zircons have notably similar and episodic appearances in the rock record. Their fluctuations correlate with Precambrian supercontinent cycles. The U–Pb zircon dates from orogenic granites are of the most reliable aging determinants. Some issues exist with relying on granite sourced zircons, such as a lack of evenly globally sourced data and the loss of granite zircons by sedimentary coverage or plutonic consumption. Where granite zircons are less adequate, detrital zircons from sandstones appear and make up for the gaps. These detrital zircons are taken from the sands of major modern rivers and their drainage basins. Oceanic magnetic anomalies and paleomagnetic data are the primary resources used for reconstructing continent and supercontinent locations back to roughly 150 Ma.

Supercontinents and Atmospheric Gases

Plate tectonics and the chemical composition of the atmosphere (specifically greenhouse gases) are the two most prevailing factors present within the geologic time scale. Continental drift influences both cold and warm climatic episodes. Atmospheric circulation and climate are strongly influenced by the location and formation of continents and megacontinents. Therefore, continental drift influences mean global temperature.

Oxygen levels of the Archaean Eon were negligible and today they are roughly 21 percent. It is thought that the Earth's oxygen content has risen in stages: six or seven steps that are timed very closely to the development of Earth's supercontinents.

The process of Earth's increase in atmospheric oxygen content is theorized to have started with continent-continent collision of huge land masses forming supercontinents, and therefore possibly supercontinent mountain ranges (supermountains). These supermountains would have eroded, and the mass amounts of nutrients, including iron and phosphorus, would have washed into oceans, just as we see happening today. The oceans would then be rich in nutrients essential to photosynthetic organisms, which would then be able to respire mass amounts of oxygen. (1: continents collide, 2: 'supermountains' form, 3: erosion of 'supermountains,' 4: large quantities of minerals and nutrients washed out to open ocean, 5: explosion of marine algae life (partly sourced from noted nutrients), and 6: mass amounts of oxygen produced during photosynthesis. There is an apparent direct relationship between orogeny and the atmospheric oxygen content). There is also evidence for increased sedimentation concurrent with the timing of these mass oxygenation events, meaning that the organic carbon and pyrite at these times were more likely to be buried beneath sediment and therefore unable to react with the free oxygen. This sustained the atmospheric oxygen increases.

During this time, 2.65 Ga there was an increase in molybdenum isotope fractionation. It was temporary, but supports the increase in atmospheric oxygen because molybdenum isotopes require free oxygen to fractionate. Between 2.45 and 2.32 Ga, the second period of oxygenation occurred, it has been called the 'great oxygenation event.' There are many pieces of evidence that support the existence of this event, including red beds appearance 2.3 Ga (meaning that Fe^{3+} was being produced and became an important component in soils). The third oxygenation stage approximately 1.8 Ga is indicated by the disappearance of iron formations. Neodymium isotopic studies suggest that iron formations are usually from continental sources, meaning that dissolved Fe and Fe^{2+} had to be transported during continental erosion. A rise in atmospheric oxygen prevents Fe transport, so the lack of iron formations may have been due to an increase in oxygen. The fourth oxygenation event, roughly 0.6 Ga, is based on modeled rates of sulfur isotopes from marine carbonate-associated sulfates. An increase (near doubled concentration) of sulfur isotopes, which is suggested by these models, would require an increase in oxygen content of the deep oceans. Between 650 and 550 Ma there were three increases in ocean oxygen levels, this period is the fifth oxygenation stage. One of the reasons indicating this period to be an oxygenation event is the increase in redox-sensitive Mo in black shales. The sixth event occurred between 360 and 260 Ma and was identified by models suggesting shifts in the balance of ^{34}S in sulfates and ^{13}C in carbonates, which were strongly influenced by an increase in atmospheric oxygen.

Supercontinent Cycle

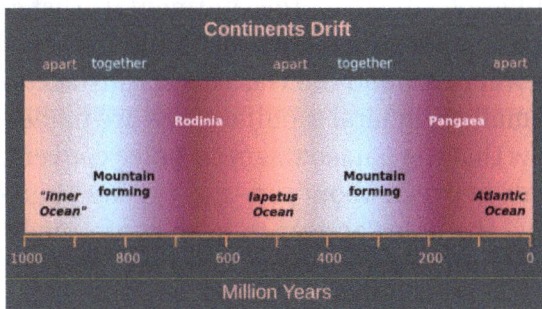

Wilson cycle

Map of Pangaea with modern continental outlines.

The supercontinent cycle is the quasi-periodic aggregation and dispersal of Earth's continental crust. There are varying opinions as to whether the amount of continental crust is increasing, decreasing, or staying about the same, but it is agreed that the Earth's crust is constantly being reconfigured. One complete supercontinent cycle is said to take 300 to 500 million years. Continental collision makes fewer and larger continents while rifting makes more and smaller continents.

Description

The most recent supercontinent, Pangaea, formed about 300 million years ago. There are two different views on the history of earlier supercontinents. The first proposes a series of supercontinents: Vaalbara (c. 3.6 to c. 2.8 billion years ago); Ur (c. 3 billion years ago); Kenorland (c. 2.7 to 2.1 billion years ago); Columbia (c. 1.8 to 1.5 billion years ago); Rodinia (c. 1.25 billion to 750 million years ago); and Pannotia (c. 600 million years ago), whose dispersal produced the fragments that ultimately collided to form Pangaea.

The second view (Protopangea-Paleopangea), based on both palaeomagnetic and geological evidence, is that supercontinent cycles did not occur before about 0.6 Ga (during the Ediacaran Period). Instead, the continental crust comprised a single supercontinent from about 2.7 Ga (Gigaanum, or "billion years ago") until it broke up for the first time, somewhere around 0.6 Ga. This reconstruction is based on the observation that if only small peripheral modifications are made to the primary reconstruction, the data show that the palaeomagnetic poles converged to quasi-static positions for long intervals between about 2.7–2.2, 1.5–1.25 and 0.75–0.6 Ga. During the intervening periods, the poles appear to have conformed to a unified apparent polar wander path. Thus the paleomagnetic data are adequately explained by the existence of a single Protopangea–Paleopangea supercontinent with prolonged quasi-integrity. The prolonged duration of this supercontinent could be explained by the operation of *lid tectonics* (comparable to the tectonics operating on Mars and Venus) during Precambrian times, as opposed to the plate tectonics seen on the contemporary Earth.

The kinds of minerals found inside ancient diamonds suggest that the cycle of super-continental formation and breakup began roughly 3.0 billion years ago. Before 3.2 billion years ago only diamonds with peridotitic compositions (commonly found in the Earth's mantle) formed, whereas after 3.0 billion years ago eclogitic diamonds (rocks from the Earth's surface crust) became prevalent. This change is thought to have come about because the process of subduction and continental collision introduced eclogite into subcontinental diamond-forming fluids.

The hypothesized supercontinent cycle is complemented by the Wilson cycle named after plate tectonics pioneer J. Tuzo Wilson, which describes the periodic opening and closing of ocean basins. Because the oldest seafloor material found today dates to only 170 million years old, whereas the oldest continental crust material found today dates to at least 4 billion years old, it makes sense to emphasize the much longer record of the planetary pulse that is recorded in the continents.

Effects on Sea Level

It is known that sea level is generally low when the continents are together and high when they are apart. For example, sea level was low at the time of formation of Pangaea (Permian) and Pannotia (latest Neoproterozoic), and rose rapidly to maxima during Ordovician and Cretaceous times, when the continents were dispersed. This is because the age of the oceanic lithosphere provides a major control on the depth of the ocean basins, and therefore on global sea level. Oceanic lithosphere forms at mid-ocean ridges and moves outwards. As this happens, it conductively cools and shrinks. This cooling and shrinking decreases the thickness and increases the density of the oceanic lithosphere, and the result is the general lowering in elevation of the seafloor away from mid-ocean ridges. For oceanic lithosphere that is less than about 75 million years old, a simple cooling half-space model of conductive cooling works, in which the depth of the ocean basins d in areas in which there is no nearby subduction is a function of the age of the oceanic lithosphere t. In general,

$$d(t) = \frac{2}{\sqrt{\pi}} a_{eff} T_1 \sqrt{\kappa t} + d_r$$

where κ is the thermal diffusivity of the mantle lithosphere (c. 8×10^{-7} m²/s), a_{eff} is the effective thermal expansion coefficient for rock (c. 5.7×10^{-5} °C^{-1}), T_1 is the temperature of ascending magma compared to the temperature at the upper boundary (c. 1220 °C for the Atlantic and Indian Oceans, c. 1120 °C for the eastern Pacific) and d_r is the depth of the ridge below the ocean surface. After plugging in rough numbers for the sea floor, the equation becomes:

for the eastern Pacific Ocean:

$$d(t) = 350\sqrt{t} + 2500$$

and for the Atlantic and Indian Oceans:

$$d(t) = 390\sqrt{t} + 2500$$

where d is in meters and t is in millions of years, so that just-formed crust at the mid-ocean ridges lies at about 2,500 m depth, whereas 50-million-year-old seafloor lies at a depth of about 5,000 m.

As the mean level of the sea floor decreases, the volume of the ocean basins increases, and if other factors that can control sea level remain constant, sea level falls. The converse is also true: younger oceanic lithosphere leads to shallower oceans and higher sea levels if other factors remain constant.

Area A, of the oceans, can change when continents rift (stretching the continents decreases A and raises sea level) or as a result of continental collision (compressing the continents leads to an increase A and lowers sea level). Increasing sea level will flood the continents, while decreasing sea level will expose continental shelves.

Because the continental shelf has a very low slope, a small increase in sea level will result in a large change in the percent of continents flooded.

If the world ocean on average is young, the seafloor will be relatively shallow, and sea level will be high: more of the continents are flooded. If the world ocean is on average old, seafloor will be relatively deep, and sea level will be low: more of the continents will be exposed.

There is thus a relatively simple relationship between the supercontinent cycle and the mean age of the seafloor.

- Supercontinent = lots of old seafloor = low sea level
- Dispersed continents = lots of young seafloor = high sea level

There will also be a climatic effect of the supercontinent cycle that will amplify this further:

- Supercontinent = continental climate dominant = continental glaciation likely = still lower sea level
- Dispersed continents = maritime climate dominant = continental glaciation unlikely = sea level is not lowered by this mechanism

Relation to Global Tectonics

There is a progression of tectonic regimes that accompany the supercontinent cycle:

During break-up of the supercontinent, rifting environments dominate. This is followed by passive margin environments, while seafloor spreading continues and the oceans grow. This in turn is followed by the development of collisional environments that become increasingly important with time. First collisions are between continents and island arcs, but lead ultimately to continent-continent collisions. This is the situation that was observed during the Paleozoic supercontinent cycle and is being observed for the Mesozoic–Cenozoic supercontinent cycle, still in progress.

Relation to Climate

There are two types of global earth climates: icehouse and greenhouse. Icehouse is characterized by frequent continental glaciations and severe desert environments. Greenhouse is characterized by warm climates. Both reflect the supercontinent cycle. We are now in a little greenhouse phase of an icehouse world.

- Icehouse climate
    - o   Continents moving together
    - o   Sea level low due to lack of seafloor production
    - o   Climate cooler, arid
    - o   Associated with aragonite seas
    - o   Formation of supercontinents
- Greenhouse climate
    - o   Continents dispersed
    - o   Sea level high
    - o   High level of sea floor spreading
    - o   Relatively large amounts of CO_2 production at oceanic rifting zones
    - o   Climate warm and humid
    - o   Associated with calcite seas

Periods of icehouse climate: much of Neoproterozoic, late Paleozoic, late Cenozoic.

Periods of greenhouse climate: Early Paleozoic, Mesozoic–early Cenozoic.

Relation to Evolution

The principal mechanism for evolution is natural selection among diverse populations. As genetic drift occurs more frequently in small populations, diversity is an observed consequence of isolation. Less isolation, and thus less diversification, occurs when the continents are all together, producing both one continent and one ocean with one coast. In Latest Neoproterozoic to Early Paleozoic times, when the tremendous proliferation of diverse metazoa occurred, isolation of marine environments resulted from the breakup of Pannotia.

A north–south arrangement of continents and oceans leads to much more diversity and isolation than east–west arrangements. North-to-south arrangements give climatically different zones along the communication routes to the north and south, which are separated by water or land from other continental or oceanic zones of similar climate. Formation of similar tracts of continents and ocean basins oriented east–west would lead to much less isolation, diversification, and slower evolution, since each continent or ocean is in fewer climatic zones. Through the Cenozoic, isolation has been maximized by a north–south arrangement.

Diversity, as measured by the number of families, follows the supercontinent cycle very well.

Gondwana

In paleogeography, Gondwana, also Gondwanaland, is the name given to an ancient supercontinent. It is believed to have sutured between about 570 and 510 million years ago (Mya), joining East Gondwana to West Gondwana. Gondwana formed prior to Pangaea, and later became part of it.

Around 300 Mya Gondwana and Laurasia joined together to form the supercontinent Pangaea, which existed until approximately 200-180 Mya. Gondwana then separated from Laurasia (the mid-Mesozoic era) in the breakup of Pangaea, drifting farther south after the split. Gondwana itself then also broke apart.

Gondwana included most of the landmasses in today's Southern Hemisphere, including Antarctica, South America, Africa, Madagascar, and the Australian continent, as well as the Arabian Peninsula and the Indian Subcontinent, which have now moved entirely into the Northern Hemisphere.

The continent of Gondwana was named by Austrian scientist Eduard Suess, after the Gondwana region of central northern India which is derived from Sanskrit for "forest of the Gonds". The name had been previously used in a geological context, first by H.B. Medlicott in 1872. from which the Gondwana sedimentary sequences (Permian-Triassic) are also described.

The adjective "Gondwanan" is in common use in biogeography when referring to patterns of distribution of living organisms, typically when the organisms are restricted to two or more of the now-discontinuous regions that were once part of Gondwana, including the Antarctic flora. For example, the Proteaceae family of plants known only from southern South America, South Africa, Australia, and New Zealand is considered to have a "Gondwanan distribution". This pattern is often considered to indicate an archaic, or relict, lineage.

Formation

The assembly of Gondwana was a protracted process. Several orogenies led to its final amalgamation 550–500 Mya at the end of the Ediacaran, and into the Cambrian. These include the Brasiliano Orogeny, the East African Orogeny, the Malagasy Orogeny, and the Kuunga Orogeny. The final stages of Gondwanan assembly overlapped with the opening of the Iapetus Ocean between Laurentia and western Gondwana. During this interval, the Cambrian explosion occurred.

Gondwana was formed from the following earlier continents and microcontinents, among others, colliding in the following orogenies:

- Azania: much of central Madagascar, the Horn of Africa and parts of Yemen and Arabia (Named by Collins and Pisarevsky (2005): "Azania" was a Greek name for the East African coast.)

- The Congo–Tanzania–Bangweulu Block of central Africa;

- Neoproterozoic India: India, the Antongil Block in far eastern Madagascar, the Seychelles, and the Napier and Rayner Complexes in East Antarctica

- The Australia/Mawson continent: Australia west of Adelaide and a large extension into East Antarctica

- Other blocks which helped to form Argentina and some surrounding regions, including a piece transferred from Laurentia when the west edge of Gondwana scraped against southeast Laurentia in the Ordovician. This is the Famatinian block (named after Famatina in northwest Argentina) and it formerly continued the line of the Appalachians southwards.

Reconstruction showing final stages of assembly of Gondwana, 550 Mya

One of the major sites of Gondwanan amalgamation was the East African Orogeny (Stern, 1994), where these two major orogenies are superimposed. The East African Orogeny at about 650–630 Mya affected a large part of Arabia, north-eastern Africa, East Africa, and Madagascar. Collins and Windley (2002) propose that in this orogeny, Azania collided with the Congo–Tanzania–Bangweulu Block.

The later Malagasy orogeny at about 550–515 Mya affected Madagascar, eastern East Africa and southern India. In it, Neoproterozoic India collided with the already combined Azania and Congo–Tanzania–Bangweulu Block, suturing along the Mozambique Belt.

At the same time, in the Kunga Orogeny Neoproterozoic India collided with the Australia/Mawson continent.

Pangaea

Other large continental masses, including the core cratons of North America (the Canadian Shield or Laurentia), Europe (Baltica), and Siberia, were added over time to form the supercontinent Pangaea by Permian time. When Pangaea broke up (mostly during the Jurassic), two large masses, Gondwana and Laurasia, were formed.

The reformed Gondwanan continent was not precisely the same as that which had existed before Pangaea formed; for example, most of Florida and southern Georgia and Alabama is underlain by rocks that were originally part of Gondwana, but that were left attached to North America when Pangaea broke apart.

Climate

During the late Paleozoic, Gondwana extended from a point at or near the South Pole to near the Equator. Across much of Gondwana, the climate was mild. During the Mesozoic, the world was on average considerably warmer than it is today. Gondwana was then host to a huge variety of flora and fauna for many millions of years. The laurel forest of Australia, New Caledonia, and New Zealand have a number of other related species of the laurissilva de Valdivia, through the connection of the Antarctic flora as gymnosperms and deciduous angiosperm *Nothofagus*. *Corynocarpus laevigatus* is called the bay of New Zealand, *Laurelia novae-zelandiae* belongs to the same genus *Laurelia*. The sempervirens tree niaouli grows in Australia, New Caledonia, and New Zealand.

New Caledonia and New Zealand ecoregions became separated from Australia by continental drift 85 million years ago. The islands still retain plants that originated in Gondwana and spread to the Southern Hemisphere continents later. However, strong evidence exists of glaciation during the Carboniferous to Permian time, especially in South Africa.

Breakup
Mesozoic

The *Nothofagus* plant genus illustrates Gondwanan distribution, having descended from the supercontinent and existing in present-day Australia, New Zealand, New Caledonia, and the Southern Cone. Fossils
have also recently been found in Antarctica.

Gondwana began to break up in the early Jurassic (about 184 Mya) accompanied by
massive eruptions of basalt lava, as East Gondwana, comprising Antarctica, Madagascar, India, and Australia, began to separate from Africa. South America began to
drift slowly westward from Africa as the South Atlantic Ocean opened, beginning about
130 Mya during the Early Cretaceous, and resulting in open marine conditions by
110 Mya. East Gondwana then began to separate about 120 Mya when India began to
move northward.

The Madagascar block, and a narrow remnant microcontinent presently occupied by
the Seychelles Islands, were broken off India; elements of this breakup nearly coincide
with the Cretaceous–Paleogene extinction event. The India–Madagascar–Seychelles
separations appear to coincide with the eruption of the Deccan basalts, whose eruption
site may survive as the Réunion hotspot.

Australia began to separate from Antarctica perhaps 80 Mya (Late Cretaceous), but
sea-floor spreading between them became most active about 40 Mya during the Eocene
epoch of the Paleogene Period.

New Zealand probably separated from Antarctica between 130 and 85 Mya.

Cenozoic

As the age of mammals commenced, the continent of Australia-New Guinea began
gradually to separate and move north (55 Mya), rotating about its axis to begin with,
and thus retaining some connection with the remainder of Gondwana for about 10 million years.

About 45 Mya, the Indian Plate collided with Asia, buckling the crust and forming the
Himalayas. At about the same time, the southernmost part of Australia (modern Tasmania) finally separated from Antarctica, letting ocean currents flow between the two

continents for the first time. Antarctica became cooler and Australia became drier because ocean currents circling Antarctica were no longer directed around northern Australia into the subtropics.

The separation of South America from West Antarctica some time during the Oligocene, perhaps 30 Mya, also caused climate changes. Immediately before this separation, South America and East Antarctica were not connected directly. However, the many microplates of the Antarctic Peninsula remained near southern South America, acting as "stepping stones" and allowing continued biological interchange and stopped oceanic current circulation. When the Drake Passage opened, a barrier was no longer present to force the cold waters of the Southern Ocean to be exchanged with warmer tropical water. Instead, a cold circumpolar current developed and Antarctica became what it is today: a frigid continent that locks up much of the world's fresh water as ice. Sea temperatures dropped by almost 10°C, and the global climate became much colder.

By about 15 Mya, the collision between New Guinea (on the leading edge of the Australian Plate) and the southwestern part of the Pacific Plate pushed up the New Guinea Highlands, causing a rain shadow effect which drastically changed weather patterns in Australia, drying it out.

Later, South America was connected to North America via the Isthmus of Panama, cutting off a circulation of warm water and thereby making the Arctic colder, as well as allowing the Great American Interchange.

The Red Sea and East African Rift are modern examples of continental rifting.

Laurasia

Laurasia was the more northern of two supercontinents (the other being Gondwana) that formed part of the Pangaea supercontinent around 300 to 200 million years ago (Mya). It separated from Gondwana 200 to 180 Mya (beginning in the late Triassic period) during the breakup of Pangaea, drifting farther north after the split.

The name combines the names of Laurentia, the name given to the North American craton, and Eurasia. As suggested by the geologic naming, Laurasia included most of the land masses which make up today's continents of the Northern Hemisphere, chiefly Laurentia, Baltica, Siberia, Kazakhstania, and the North China and East China cratons.

Origin

Although Laurasia is known as a Mesozoic phenomenon, today it is believed that the same continents that formed the later Laurasia also existed as a coherent supercontinent after the breakup of Rodinia around 1 billion years ago. To avoid confusion with the Mesozoic continent, this is referred to as Proto-Laurasia. It is believed that Laurasia did not break up again before it recombined with the southern continents to form the late Precambrian supercontinent of Pannotia, which remained until the early Cambrian. Laurasia was assembled, then broken up, due to the actions of plate tectonics, continental drift, and seafloor spreading.

Breakup and Reformation

During the Cambrian, Laurasia was largely located in equatorial latitudes and began to break up, with North China and Siberia drifting into latitudes further north than those occupied by continents during the previous 500 million years. By the Devonian, North China was located near the Arctic Circle and it remained the northernmost land in the world during the Carboniferous Ice Age between 300 and 280 million years ago. No evidence, though, exists for any large-scale Carboniferous glaciation of the northern continents. This cold period saw the rejoining of Laurentia and Baltica with the formation of the Appalachian Mountains and the vast coal deposits, which are a mainstay of the economies of such regions as West Virginia, Britain, and Germany.

Siberia moved southwards and joined with Kazakhstania, a small continental region believed today to have been created during the Silurian by extensive volcanism. When these two continents joined together, Laurasia was nearly reformed, and by the beginning of the Triassic, the East China craton had rejoined the redeveloping Laurasia as it collided with Gondwana to form Pangaea. North China became, as it drifted southwards from near-Arctic latitudes, the last continent to join with Pangaea.

Final Split

Around 200 million years ago, Pangaea started to break up. Between eastern North America and northwest Africa, a new ocean formed - the Atlantic Ocean, though Greenland (attached to North America) and Europe were still joined together. The separation of Europe and Greenland occurred around 55 million years ago (at the end of the Paleocene). Laurasia finally divided into the continents after which it is named: Laurentia (now North America) and Eurasia (excluding the Indian subcontinent).

Pangaea

Pangaea or Pangea was a supercontinent that existed during the late Paleozoic and early Mesozoic eras. It assembled from earlier continental units approximately 300 million years ago, and it began to break apart about 175 million years ago. In contrast to the present Earth and its distribution of continental mass, much of Pangaea was in the southern hemisphere and surrounded by a superocean, Panthalassa. Pangaea was the last supercontinent to have existed and the first to be reconstructed by geologists.

Origin of the Concept

The name "Pangaea" is derived from Ancient Greek *pan* and *Gaia*. The concept that the continents once formed a continuous land mass was first proposed by Alfred Wegener, the originator of the theory of continental drift, in his 1912 publication *The Origin of Continents* (*Die Entstehung der Kontinente*). He expanded upon his hypothesis in his 1915 book *The Origin of Continents and Oceans* (*Die Entstehung der Kontinente und Ozeane*), in which he postulated that, before breaking up and drifting to their present locations, all the continents had formed a single supercontinent that he called the "*Urkontinent*". The name "Pangea" occurs in the 1920 edition of *Die Entstehung der Kontinente und Ozeane*, but only once, when Wegener refers to the ancient supercontinent as "the Pangaea of the Carboniferous". Wegener used the Germanized form "Pangäa", but the name entered German and English scientific literature (in 1922 and 1926, respectively) in the Latinized form "Pangaea" (of the Greek "Pangaia"), especially due to a symposium of the American Association of Petroleum Geologists in November 1926.

Formation

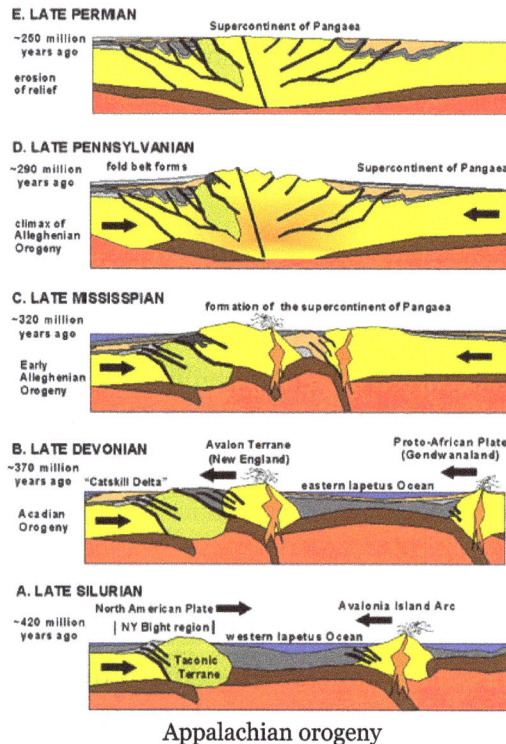

Appalachian orogeny

The forming of supercontinents and their breaking up appears to have been cyclical through Earth's history. There may have been many others before Pangaea. The fourth-last supercontinent, called Columbia or Nuna, appears to have assembled in the period 2.0–1.8 Ga. Columbia/Nuna broke up and the next supercontinent, Rodinia, formed from the accretion and assembly of its fragments. Rodinia lasted from about 1.1 billion years ago (Ga) until about 750 million years ago, but its exact configuration and geodynamic history are not nearly as well understood as those of the later supercontinents, Pannotia and Pangaea.

When Rodinia broke up, it split into three pieces: the supercontinent of Proto-Laurasia, the supercontinent of Proto-Gondwana, and the smaller Congo craton. Proto-Laurasia and Proto-Gondwana were separated by the Proto-Tethys Ocean. Next Proto-Laurasia itself split apart to form the continents of Laurentia, Siberia and Baltica. Baltica moved to the east of Laurentia, and Siberia moved northeast of Laurentia. The splitting also created two new oceans, the Iapetus Ocean and Paleoasian Ocean. Most of the above masses coalesced again to form the relatively short-lived supercontinent of Pannotia. This supercontinent included large amounts of land near the poles and, near the equator, only a relatively small strip connecting the polar masses. Pannotia lasted until 540 Ma, near the beginning of the Cambrian period and then broke up, giving rise to the continents of Laurentia, Baltica, and the southern supercontinent of Gondwana.

In the Cambrian period, the continent of Laurentia, which would later become North America, sat on the equator, with three bordering oceans: the Panthalassic Ocean to the north and west, the Iapetus Ocean to the south and the Khanty Ocean to the east. In the Earliest Ordovician, around 480 Ma, the microcontinent of Avalonia – a landmass incorporating fragments of what would become eastern Newfoundland, the southern British Isles, and parts of Belgium, northern France, Nova Scotia, New England, Iberia and northwest Africa – broke free from Gondwana and began its journey to Laurentia. Baltica, Laurentia, and Avalonia all came together by the end of the Ordovician to form a minor supercontinent called Euramerica or Laurussia, closing the Iapetus Ocean. The collision also resulted in the formation of the northern Appalachians. Siberia sat near Euramerica, with the Khanty Ocean between the two continents. While all this was happening, Gondwana drifted slowly towards the South Pole. This was the first step of the formation of Pangaea.

The second step in the formation of Pangaea was the collision of Gondwana with Euramerica. By Silurian time, 440 Ma, Baltica had already collided with Laurentia, forming Euramerica. Avalonia had not yet collided with Laurentia, but as Avalonia inched towards Laurentia, the seaway between them, a remnant of the Iapetus Ocean, was slowly shrinking. Meanwhile, southern Europe broke off from Gondwana and began to move towards Euramerica across the newly formed Rheic Ocean. It collided with southern Baltica in the Devonian, though this microcontinent was an underwater plate. The Iapetus Ocean's sister ocean, the Khanty Ocean, shrank as an island arc from Siberia collided with eastern Baltica (now part of Euramerica). Behind this island arc was a new ocean, the Ural Ocean.

By late Silurian time, North and South China split from Gondwana and started to head northward, shrinking the Proto-Tethys Ocean in their path and opening the new Paleo-Tethys Ocean to their south. In the Devonian Period, Gondwana itself headed towards Euramerica, causing the Rheic Ocean to shrink. In the Early Carboniferous, northwest Africa had touched the southeastern coast of Euramerica, creating the southern portion of the Appalachian Mountains, the Meseta Mountains and the Mauritanide Mountains. South America moved northward to southern Euramerica, while the eastern portion of Gondwana (India, Antarctica and Australia) headed toward the South Pole from the equator. North and South China were on independent continents. The Kazakhstania microcontinent had collided with Siberia. (Siberia had been a separate continent for millions of years since the deformation of the supercontinent Pannotia in the Middle Carboniferous.)

Western Kazakhstania collided with Baltica in the Late Carboniferous, closing the Ural Ocean between them and the western Proto-Tethys in them (Uralian orogeny), causing the formation of not only the Ural Mountains but also the supercontinent of Laurasia. This was the last step of the formation of Pangaea. Meanwhile, South America had collided with southern Laurentia, closing the Rheic Ocean and forming the southernmost part of the Appalachians and Ouachita Mountains. By this time, Gondwana was positioned near the South Pole and glaciers were forming in Antarctica, India, Australia, southern Africa and South America. The North China block collided with Siberia by Late Carboniferous time, completely closing the Proto-Tethys Ocean.

By Early Permian time, the Cimmerian plate split from Gondwana and headed towards Laurasia, thus closing the Paleo-Tethys Ocean, but forming a new ocean, the Tethys Ocean, in its southern end. Most of the landmasses were all in one. By the Triassic Period, Pangaea rotated a little and the Cimmerian plate was still travelling across the shrinking Paleo-Tethys, until the Middle Jurassic time. The Paleo-Tethys had closed from west to east, creating the Cimmerian Orogeny. Pangaea, which looked like a *C*, with the new Tethys Ocean inside the *C*, had rifted by the Middle Jurassic, and its deformation is explained below.

Evidence of Existence

Fossil evidence for Pangaea includes the presence of similar and identical species on continents that are now great distances apart. For example, fossils of the therapsid *Lystrosaurus* have been found in South Africa, India and Antarctica, alongside members of the *Glossopteris* flora, whose distribution would have ranged from the polar circle to the equator if the continents had been in their present position; similarly, the freshwater reptile *Mesosaurus* has been found in only localized regions of the coasts of Brazil and West Africa.

Additional evidence for Pangaea is found in the geology of adjacent continents, including matching geological trends between the eastern coast of South America and the western coast of Africa. The polar ice cap of the Carboniferous Period covered the southern end of Pangaea. Glacial deposits, specifically till, of the same age and structure are found on many separate continents which would have been together in the continent of Pangaea.

Paleomagnetic study of apparent polar wandering paths also support the theory of a supercontinent. Geologists can determine the movement of continental plates by examining the orientation of magnetic minerals in rocks; when rocks are formed, they take on the magnetic properties of the Earth and indicate in which direction the poles lie relative to the rock. Since the magnetic poles drift about the rotational pole with a period of only a few thousand years, measurements from numerous lavas spanning several thousand years are averaged to give an apparent mean polar position. Samples of sedimentary rock and intrusive igneous rock have magnetic orientations that are typically an average of the "secular variation" in the orientation of magnetic north because their remanent magnetizations are not acquired instantaneously. Magnetic differences between sample groups whose age varies by millions of years is due to a combination of true polar wander and the drifting of continents. The true polar wander component is identical for all samples, and can be removed, leaving geologists with the portion of this motion that shows continental drift and can be used to help reconstruct earlier continental positions.

The continuity of mountain chains provides further evidence for Pangaea. One example of this is the Appalachian Mountains chain which extends from the southeastern United States to the Caledonides of Ireland, Britain, Greenland, and Scandinavia.

Rifting and Break-up

There were three major phases in the break-up of Pangaea. The first phase began in the Early-Middle Jurassic (about 175 Ma), when Pangaea began to rift from the Tethys Ocean in the east to the Pacific in the west. The rifting that took place between North America and Africa produced multiple failed rifts. One rift resulted in a new ocean, the North Atlantic Ocean.

The Atlantic Ocean did not open uniformly; rifting began in the north-central Atlantic. The South Atlantic did not open until the Cretaceous when Laurasia started to rotate clockwise and moved northward with North America to the north, and Eurasia to the south. The clockwise motion of Laurasia led much later to the closing of the Tethys Ocean. Meanwhile, on the other side of Africa and along the adjacent margins of east Africa, Antarctica and Madagascar, new rifts were forming that would lead to the formation of the southwestern Indian Ocean that would open up in the Cretaceous.

The second major phase in the break-up of Pangaea began in the Early Cretaceous (150–140 Ma), when the minor supercontinent of Gondwana separated into multiple continents (Africa, South America, India, Antarctica, and Australia). The subduction at

Tethyan Trench probably caused Africa, India and Australia to move northward, caus-ing the opening of a "South Indian Ocean". In the Early Cretaceous, Atlantica, today's South America and Africa, finally separated from eastern Gondwana (Antarctica, India and Australia). Then in the Middle Cretaceous, Gondwana fragmented to open up the South Atlantic Ocean as South America started to move westward away from Africa. The South Atlantic did not develop uniformly; rather, it rifted from south to north.

Also, at the same time, Madagascar and India began to separate from Antarctica and moved northward, opening up the Indian Ocean. Madagascar and India separated from each other 100–90 Ma in the Late Cretaceous. India continued to move north-ward toward Eurasia at 15 centimeters (6 in) a year (a plate tectonic record), closing the eastern Tethys Ocean, while Madagascar stopped and became locked to the Afri-can Plate. New Zealand, New Caledonia and the rest of Zealandia began to separate from Australia, moving eastward toward the Pacific and opening the Coral Sea and Tasman Sea.

The third major and final phase of the break-up of Pangaea occurred in the early Cenozoic (Paleocene to Oligocene). Laurasia split when North America/Greenland (also called Laurentia) broke free from Eurasia, opening the Norwegian Sea about 60–55 Ma. The Atlantic and Indian Oceans continued to expand, closing the Tethys Ocean.

Meanwhile, Australia split from Antarctica and moved quickly northward, just as In-dia had done more than 40 million years before. Australia is currently on a collision course with eastern Asia. Both Australia and India are currently moving northeast at 5–6 centimeters (2–3 in) a year. Antarctica has been near or at the South Pole since the formation of Pangaea about 280 Ma. India started to collide with Asia be-ginning about 35 Ma, forming the Himalayan orogeny, and also finally closing the Tethys Seaway; this collision continues today. The African Plate started to change di-rections, from west to northwest toward Europe, and South America began to move in a northward direction, separating it from Antarctica and allowing complete oceanic circulation around Antarctica for the first time. This motion, together with decreasing atmospheric carbon dioxide concentrations, caused a rapid cooling of Antarctica and allowed glaciers to form. This glaciation eventually coalesced into the kilometers-thick ice sheets seen today. Other major events took place during the Cenozoic, including the opening of the Gulf of California, the uplift of the Alps, and the opening of the Sea of Japan. The break-up of Pangaea continues today in the Red Sea Rift and East Afri-can Rift.

Tectonic Plate Shift

Pangaea's formation is now commonly explained in terms of plate tectonics. The in-volvement of plate tectonics in Pangaea's separation helps to show how it did not sep-arate all at once, but at different times, in sequences. Additionally, after these separations, it has also been discovered that the separated land masses may have also continued to break apart multiple times. The formation of each environment and climate on Pangaea is due to plate tectonics, and thus, it is as a result of these shifts and changes different climatic pressures were placed on the life on Pangaea. Although plate tectonics was paramount in the formation of later land masses, it was also essential in the placement, climate, environments, habitats, and overall structure of Pangaea.

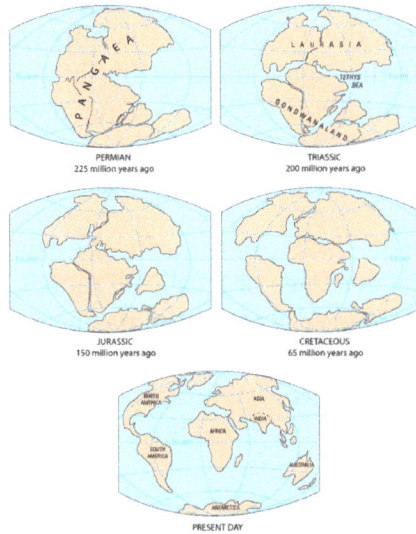

The breakup of Pangaea over time

What can also be observed in relation to tectonic plates and Pangaea, is the formations to such plates. Mountains and valleys form due to tectonic collisions as well as earthquakes and chasms. Consequentially, this shaped Pangaea and animal adaptations. Furthermore, plate tectonics can contribute to volcanic activity, which is responsible for extinctions and adaptations which have evidently affected life over time, and without doubt on Pangaea.

Life

Example of an Ammonite

Over the 100 million years Pangaea existed, many species had fruitful times whereas others struggled. The Traversodontidae is an example of such prospering animals, eating a diet of only plants. Plants dependent on spore reproduction had been taken out of the ecosystems, and replaced by the gymnosperm plant, which reproduces through the use of seeds instead. These plants were also able to transport water internally, allowing animals that ate it to also improve hydration. Later on, insects (beetles, dragonflies, mosquitos) also thrived during the Permian period (250–300) million years ago. However, the Permian extinction would eventually come and greatly impact these insects through a mass extinction, being the only mass extinction to affect insects. When the Triassic Period came, many reptiles were able to also thrive, including Archosaurs, which is an ancestor to modern-day crocodiles and birds.

Little is known about marine life back to date during the existence of Pangaea. Scientists are unable to find substantial evidence or fossilized remains in order to assist them in answering such a question. However, a couple of marine animals have been discovered to exist at the time- the Ammonites and Brachiopods. Additionally, evidence pointing towards massive reefs with varied ecosystems, especially in the species of sponges and coral, has also been discovered.

Climate Change After Pangaea

Pangaea has tremendously affected the setup of the world now. We live in a post Pangaea time period where the reconfiguration of continents and oceans has changed the climate of many areas. There is scientific evidence that proves that climate was drastically altered. When the continents separated and reformed themselves, it changed the flow of the oceanic currents and winds. The scientific reasoning behind all of the changes is Continental Drift. The theory of Continental Drift, created by Alfred Wegener, explained how the continents shifted Earth's surface and how that affected many aspects such as climate, rock formations found on different continents and plant and animal fossils. Wegener studied plant fossils from the frigid Arctic of Svalbard, Norway. He determined that such plants were not meant to adapt to a glacial climate. The fossils he found were from tropical plants that were meant to adapt and thrive in warmer and tropical climate. Because we would not assume that the plant fossils were capable of traveling to a different place we suspect that Svalbard possibly had a warmer, less frigid climate in the past.

When Pangaea separated, the reorganization of the continents changed the function of the oceans and seaways. The restructuring of the continents, changed and altered the distribution of warmth and coolness of the oceans. When North America and South America connected, it stopped equatorial currents from passing from the Atlantic Ocean to the Pacific Ocean. Researchers have found evidence by using computer hydrological models to show that this strengthened the Gulf Stream by diverting more warm currents towards Europe. Warm waters at high latitudes led to an increased evaporation and eventually atmospheric moisture. Increased evaporation and atmospheric moisture resulted in increased precipitation. Evidence of increased precipitation is the development of snow and ice that covers Greenland, which led to an accumulation of the icecap. Greenland's growing ice cap led to further global cooling. Scientists also found evidence of global cooling through the separation of Australia and Antarctica and the formation of the Antarctic Ocean. Ocean currents in the newly formed Antarctic or Southern Ocean created a circumpolar current. The creation of the new ocean that caused a circumpolar current eventually led to atmospheric currents that rotated from west to east. Atmospheric and oceanic currents stopped the transfer of warm, tropical air and water to the higher latitudes. As a result of the warm air and currents moving northward, Antarctica cooled down so much that it became frigid.

Although many of Alfred Wegener's theories and conclusions were valid, scientists are constantly coming up with new innovative ideas or reasoning behind why certain things happen. Wegener's theory of Continental Drift was later replaced by the theory of tectonic plates.

Implications of Extinction

There is evidence to suggest that the deterioration of northern Pangaea contributed to the Permian Extinction, one of Earth's five major mass extinction events, which resulted in the loss of over 90% of marine and 70% of terrestrial species. There were three main sources of environmental deterioration which are believed to have had a hand in the extinction event.

The first of these sources is a loss of oxygen concentration in the ocean which caused deep water regions called the lysocline to grow shallower. With the lysocline shrinking, there were fewer places for calcite to dissolve in the ocean, considering calcite only dissolves at deep ocean depths. This led to the extinction of carbonate producers such as brachiopods and corals that relied on dissolved calcite to survive. The second source is the eruption of the Siberian Traps, a large volcanic event which is argued to be the result of Pangaean tectonic movement. This had several negative repercussions on the environment, including metal loading and excess atmospheric carbon. Metal loading, the release of toxic metals from volcanic eruptions into the environment, led to acid rain and general stress on the environment. These toxic metals are known to infringe on vascular plants' ability to photosynthesize, which may have resulted in the loss of Permian-era flora. Excess CO_2 in the atmosphere is believed to be the main cause of the shrinking of lysocline areas. The third cause of this extinction event that can be attributed to northern Pangaea is the beginnings of anoxic ocean environments, or oceans with very low oxygen concentrations. The mix of anoxic oceans and ocean acidification due to metal loading led to increasingly acidic oceans, which ultimately led to the extinction of benthic species.

Rodinia

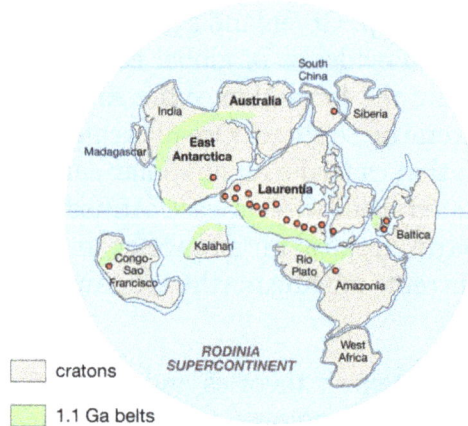

Proposed reconstruction of Rodinia for 750 Ma, with orogenic belts of 1.1 Ga age highlighted in green. Red dots indicate 1.3–1.5 Ga A-type granites.

Rodinia (from the Russian "Родина", *ródina*, meaning "The Motherland") is a Neoproterozoic supercontinent that was assembled 1.3–0.9 billion years ago and broke up 750–600 million years ago. Valentine & Moores 1970 were probably the first to recognise a Precambrian supercontinent, which they named 'Pangaea I'. It was renamed 'Rodinia' by McMenamin & McMenamin 1990 who also were the first to produce a reconstruction and propose a temporal framework for the supercontinent.

Rodinia formed at c. 1.0 Ga by accretion and collision of fragments produced by breakup of an older supercontinent, Columbia, assembled by global-scale 2.0–1.8 Ga collisional events.

Rodinia broke up in the Neoproterozoic with its continental fragments reassembled to form Pannotia 600–550 million years ago. In contrast with Pannotia, little is known yet about the exact configuration and geodynamic history of Rodinia. Paleomagnetic evidence provides some clues to the paleolatitude of individual pieces of the Earth's crust, but not to their longitude, which geologists have pieced together by comparing similar geologic features, often now widely dispersed.

The extreme cooling of the global climate around 700 million years ago (the so-called Snowball Earth of the Cryogenian Period) and the rapid evolution of primitive life during the subsequent Ediacaran and Cambrian periods are thought to have been triggered by the breaking up of Rodinia or to a slowing down of tectonic processes.

Geodynamics

Paleogeographic Reconstructions

The idea that a supercontinent existed in the early Neoproterozoic arose in the 1970s, when geologists determined that orogens of this age exist on virtually all cratons. Examples are the Grenville orogeny in North America and the Dalslandian orogeny in Europe.

Since then, many alternative reconstructions have been proposed for the configuration of the cratons in this supercontinent. Most of these reconstructions are based on the correlation of the orogens on different cratons. Though the configuration of the core cratons in Rodinia is now reasonably well known, recent reconstructions still differ in many details. Geologists try to decrease the uncertainties by collecting geological and paleomagnetical data.

Most reconstructions show Rodinia's core formed by the North American craton (the later paleocontinent of Laurentia), surrounded in the southeast with the East European craton (the later paleocontinent of Baltica), the Amazonian craton ("Amazonia") and the West African craton; in the south with the Río de la Plata and São Francisco cratons; in the southwest with the Congo and Kalahari cratons; and in the northeast with Australia, India and eastern Antarctica. The positions of Siberia and North and South China north of the North American craton differ strongly depending on the reconstruction:

- SWEAT-Configuration (Southwest US-East Antarctica craton): Antarctica is on the Southwest of Laurentia and Australia is at the North of Antarctica.

- AUSWUS-Configuration (Australia-western US): Australia is at the West of Laurentia.

- AUSMEX-Configuration (Australia-Mexico): Australia is at the location of nowadays Mexico relative to Laurentia.

- The "Missing-link" model by Li et al. 2008 which has South China between Australia and the west coast of Laurentia.

- Siberia attached to the western US (via the Belt Supergroup), as in Sears & Price 2000.

- Rodinia of Scotese.

Little is known about the paleogeography before the formation of Rodinia. Paleomagnetic and geologic data are only definite enough to form reconstructions from the breakup of Rodinia onwards. Rodinia is considered to have formed between 1.1 billion and 1 billion years ago and broke up again before 750 million years ago. Rodinia was surrounded by the superocean geologists are calling Mirovia (from Russian мировой, *mirovoy*, meaning "global").

According to J.D.A. Piper, Rodinia is one of two models for the configuration and history of the continental crust in the latter part of Precambrian times. The other is Paleopangea, Piper's own concept. Piper proposes an alternative hypothesis for this era and the previous ones. This idea rejects that Rodinia ever existed as a transient supercontinent subject to progressive break-up in the latter part of Proterozoic times and instead that this time and earlier times were dominated by a single, persistent "Paleopangaea" supercontinent. As evidence, he suggests an observation that the palaeomagnetic poles from the continental crust assigned to this time conform to a single path between 800 and 600 million years ago and latterly to a near-static position between 750 and 600 million years. This latter solution predicts that break-up was confined to the Ediacaran Period and produced the dramatic environmental changes that characterised the transition between Precambrian and Phanerozoic times.

Break up

In 2009 UNESCO's IGCP project 440, named 'Rodinia Assembly and Breakup', concluded that Rodinia broke-up in four stages between 825–550 Ma:

- The break-up was initiated by a superplume around 825–800 Ma whose influence — such as crustal arching, intense bimodal magmatism, and accumulation of thick rift-type sedimentary successions — have been recorded in South Australia, South China, Tarim, Kalahari, India, and the Arabian-Nubian Craton.

- Rifting progressed in the same cratons 800–750 Ma and spread into Laurentia and perhaps Siberia. India (including Madagascar) and the Congo-São Francisco Craton were either detached from Rodinia during this period or simply never were part of the supercontinent.

- As the central part of Rodinia reached the Equator around 750–700 Ma, a new pulse of magmatism and rifting continued the disassembly in western Kalahari, West Australia, South China, Tarim, and most margins of Laurentia.

- 650–550 Ma several events coincided: the opening of the Iapetus Ocean; the closure of the Braziliano, Adamastor, and Mozambique oceans; and the Pan-African orogeny. The result was the formation of Gondwana.

The Rodinia hypothesis assumes that rifting did not start everywhere simultaneously. Extensive lava flows and volcanic eruptions of Neoproterozoic age are found on most continents, evidence for large scale rifting about 750 million years ago. As early as 850 and 800 million years ago, a rift developed between the continental masses of present-day Australia, East Antarctica, India and the Congo and Kalahari cratons on one side and later Laurentia, Baltica, Amazonia and the West African and Rio de la Plata cratons on the other. This rift developed into the Adamastor Ocean during the Ediacaran.

Around 550 million years ago, on the boundary between the Ediacaran and Cambrian, the first group of cratons eventually fused again with Amazonia, West Africa and the Rio de la Plata cratons. This tectonic phase is called the Pan-African orogeny. It created a configuration of continents that would remain stable for hundreds of millions of years in the form of the continent Gondwana.

In a separate rifting event about 610 million years ago (halfway in the Ediacaran period), the Iapetus Ocean formed. The eastern part of this ocean formed between Baltica and Laurentia, the western part between Amazonia and Laurentia. Because the exact moments of this separation and the partially contemporaneous Pan-African orogeny are hard to correlate, it might be that all continental mass was again joined in one supercontinent between roughly 600 and 550 million years ago. This hypothetical supercontinent is called Pannotia.

Influence on Paleoclimate and Life

Unlike later supercontinents, Rodinia would have been entirely barren. Rodinia existed before complex life colonized dry land. Based on sedimentary rock analysis Rodinia's formation happened when the ozone layer was not as extensive as it is today. Ultraviolet light discouraged organisms from inhabiting its interior. Nevertheless, its existence did significantly influence the marine life of its time.

In the Cryogenian period the Earth experienced large glaciations, and temperatures were at least as cool as today. Substantial areas of Rodinia may have been covered by glaciers or the southern polar ice cap.

Low temperatures may have been exaggerated during the early stages of continental rifting. Geothermal heating peaks in crust about to be rifted; and since warmer rocks are less dense, the crustal rocks rise up relative to their surroundings. This rising creates areas of higher altitude, where the air is cooler and ice is less likely to melt with changes in season, and it may explain the evidence of abundant glaciation in the Ediacaran period.

The eventual rifting of the continents created new oceans and seafloor spreading, which produces warmer, less dense oceanic lithosphere. Due to its lower density, hot oceanic lithosphere will not lie as deep as old, cool oceanic lithosphere. In periods with relatively large areas of new lithosphere, the ocean floors come up, causing the eustatic sea level to rise. The result was a greater number of shallower seas.

The increased evaporation from the larger water area of the oceans may have increased rainfall, which, in turn, increased the weathering of exposed rock. By inputting data on the ratio of stable isotopes ^{18}O:^{16}O into computer models, it has been shown that, in conjunction with quick weathering of volcanic rock, this increased rainfall may have reduced greenhouse gas levels to below the threshold required to trigger the period of extreme glaciation known as Snowball Earth.

Increased volcanic activity also introduced into the marine environment biologically active nutrients, which may have played an important role in the development of the earliest animals.

References

- Rogers, J.J.W.; Santosh, M. (2004), Continents and Supercontinents, Oxford: Oxford University Press, p. 146, ISBN 0-19-516589-6

- Murck, Barbara W. and Skinner, Brian J. (1999) Geology Today: Understanding Our Planet, Study Guide, Wiley, ISBN 978-0-471-32323-5

- Kearey, Philip; Klepeis, Keith A. and Vine, Frederick J. (2009). Global Tectonics (3rd. ed), pp. 66–67. Chichester:Wiley. ISBN 978-1-4051-0777-8

- Ivanov, A. V. (2007). "Evaluation of different models for the origin of the Siberian traps". GSA Special Papers. 430: 669–691. doi:10.1130/2007.2430(31). ISBN 978-0-8137-2430-0.

- "Pangaea to the Present Lesson #2 | Volcano World | Oregon State University". volcano.oregon-state.edu. Retrieved 2015-10-29.

- Bradley, Dwight C., "Secular Trends in the Geologic Record and the Supercontinent Cycle." Earth Science Review. (2011): 1–18.

- Read, J. Fred (2001). "Record of ancient climates can be a map to riches". Science from Virginia Tech. Retrieved 2011-05-04.

- Piper, J.D.A. "Protopangea: palaeomangetic definition of Earth's oldest (Mid-Archaean-Paleoproterozoic) supercontinent." Journal of Geodynamics. 50 (2010): 154–165.

- "Other Reconstructions for Rodinia based on sources for Mojavia". Department of Geological Sciences, University of Colorado Boulder. May 2002. Retrieved 20 September 2010.

9

Earthquakes: An Integrated Study

An earthquake occurs because of the energy suddenly released by the Earth's crust. Earthquakes cause immense damage to life and to property. The topics covered in this section are seismotectonics, intraplate earthquake and interplate earthquake. The topics discussed in the chapter are of great importance to broaden the existing knowledge on Earthquakes.

Earthquake

An earthquake (also known as a quake, tremor or temblor) is the perceptible shaking of the surface of the Earth, resulting from the sudden release of energy in the Earth's crust that creates seismic waves. Earthquakes can be violent enough to toss people around and destroy whole cities. The seismicity or seismic activity of an area refers to the frequency, type and size of earthquakes experienced over a period of time.

Earthquakes are measured using observations from seismometers. The moment magnitude is the most common scale on which earthquakes larger than approximately 5 are reported for the entire globe. The more numerous earthquakes smaller than magnitude 5 reported by national seismological observatories are measured mostly on the local magnitude scale, also referred to as the Richter magnitude scale. These two scales are numerically similar over their range of validity. Magnitude 3 or lower earthquakes are mostly imperceptible or weak and magnitude 7 and over potentially cause serious damage over larger areas, depending on their depth. The largest earthquakes in historic times have been of magnitude slightly over 9, although there is no limit to the possible magnitude. Intensity of shaking is measured on the modified Mercalli scale. The shallower an earthquake, the more damage to structures it causes, all else being equal.

At the Earth's surface, earthquakes manifest themselves by shaking and sometimes displacement of the ground. When the epicenter of a large earthquake is located offshore, the seabed may be displaced sufficiently to cause a tsunami. Earthquakes can also trigger landslides, and occasionally volcanic activity.

In its most general sense, the word *earthquake* is used to describe any seismic event — whether natural or caused by humans — that generates seismic waves. Earthquakes are caused mostly by rupture of geological faults, but also by other events such as volcanic activity, landslides, mine blasts, and nuclear tests. An earthquake's point of initial rupture is called its focus or hypocenter. The epicenter is the point at ground level directly above the hypocenter.

Naturally Occurring Earthquakes

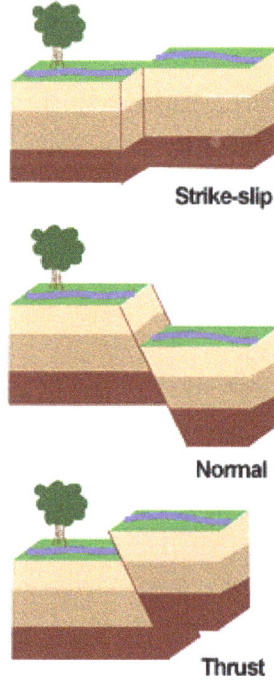

Fault types

Tectonic earthquakes occur anywhere in the earth where there is sufficient stored elastic strain energy to drive fracture propagation along a fault plane. The sides of a fault move past each other smoothly and aseismically only if there are no irregularities or asperities along the fault surface that increase the frictional resistance. Most fault surfaces do have such asperities and this leads to a form of stick-slip behavior. Once the fault has locked, continued relative motion between the plates leads to increasing stress and therefore, stored strain energy in the volume around the fault surface. This continues until the stress has risen sufficiently to break through the asperity, suddenly allowing sliding over the locked portion of the fault, releasing the stored energy. This energy is released as a combination of radiated elastic strain seismic waves, frictional heating of the fault surface, and cracking of the rock, thus causing an earthquake. This process of gradual build-up of strain and stress punctuated by occasional sudden earthquake failure is referred to as the elastic-rebound theory. It is estimated that only 10 percent or less of an earthquake's total energy is radiated as seismic energy. Most of the earthquake's energy is used to power the earthquake fracture growth or is converted into heat generated by friction. Therefore, earthquakes lower the Earth's available elastic potential energy and raise its temperature, though these changes are negligible compared to the conductive and convective flow of heat out from the Earth's deep interior.

Earthquake Fault Types

There are three main types of fault, all of which may cause an interplate earthquake: normal, reverse (thrust) and strike-slip. Normal and reverse faulting are examples of dip-slip, where the displacement along the fault is in the direction of dip and movement on them involves a vertical component. Normal faults occur mainly in areas where the crust is being extended such as a divergent boundary. Reverse faults occur in areas where the crust is being shortened such as at a convergent boundary. Strike-slip faults are steep structures where the two sides of the fault slip horizontally past each other; transform boundaries are a particular type of strike-slip fault. Many earthquakes are caused by movement on faults that have components of both dip-slip and strike-slip; this is known as oblique slip.

Reverse faults, particularly those along convergent plate boundaries are associated with the most powerful earthquakes, megathrust earthquakes, including almost all of those of magnitude 8 or more. Strike-slip faults, particularly continental transforms, can produce major earthquakes up to about magnitude 8. Earthquakes associated with normal faults are generally less than magnitude 7. For every unit increase in magnitude, there is a roughly thirtyfold increase in the energy released. For instance, an earthquake of magnitude 6.0 releases approximately 30 times more energy than a 5.0 magnitude earthquake and a 7.0 magnitude earthquake releases 900 times (30 × 30) more energy than a 5.0 magnitude of earthquake. An 8.6 magnitude earthquake releases the same amount of energy as 10,000 atomic bombs like those used in World War II.

This is so because the energy released in an earthquake, and thus its magnitude, is proportional to the area of the fault that ruptures and the stress drop. Therefore, the longer the length and the wider the width of the faulted area, the larger the resulting magnitude. The topmost, brittle part of the Earth's crust, and the cool slabs of the tectonic plates that are descending down into the hot mantle, are the only parts of our planet which can store elastic energy and release it in fault ruptures. Rocks hotter than about 300 degrees Celsius flow in response to stress; they do not rupture in earthquakes. The maximum observed lengths of ruptures and mapped faults (which may break in a single rupture) are approximately 1000 km. Examples are the earthquakes in Chile, 1960; Alaska, 1957; Sumatra, 2004, all in subduction zones. The longest earthquake ruptures on strike-slip faults, like the San Andreas Fault (1857, 1906), the North Anatolian Fault in Turkey (1939) and the Denali Fault in Alaska (2002), are about half to one third as long as the lengths along subducting plate margins, and those along normal faults are even shorter.

The most important parameter controlling the maximum earthquake magnitude on a fault is however not the maximum available length, but the available width because the latter varies by a factor of 20. Along converging plate margins, the dip angle of the rupture plane is very shallow, typically about 10 degrees. Thus the width of the plane within the top brittle crust of the Earth can become 50 to 100 km (Japan, 2011; Alaska, 1964), making the most powerful earthquakes possible.

Aerial photo of the San Andreas Fault in the Carrizo Plain, northwest of Los Angeles

Strike-slip faults tend to be oriented near vertically, resulting in an approximate width of 10 km within the brittle crust, thus earthquakes with magnitudes much larger than 8 are not possible. Maximum magnitudes along many normal faults are even more limited because many of them are located along spreading centers, as in Iceland, where the thickness of the brittle layer is only about 6 km.

In addition, there exists a hierarchy of stress level in the three fault types. Thrust faults are generated by the highest, strike slip by intermediate, and normal faults by the lowest stress levels. This can easily be understood by considering the direction of the greatest principal stress, the direction of the force that 'pushes' the rock mass during the faulting. In the case of normal faults, the rock mass is pushed down in a vertical direction, thus the pushing force (greatest principal stress) equals the weight of the rock mass itself. In the case of thrusting, the rock mass 'escapes' in the direction of the least principal stress, namely upward, lifting the rock mass up, thus the overburden equals the least principal stress. Strike-slip faulting is intermediate between the other two types described above. This difference in stress regime in the three faulting environments can contribute to differences in stress drop during faulting, which contributes to differences in the radiated energy, regardless of fault dimensions.

Earthquakes Away from Plate Boundaries

Where plate boundaries occur within the continental lithosphere, deformation is spread out over a much larger area than the plate boundary itself. In the case of the San Andreas fault continental transform, many earthquakes occur away from the plate boundary and are related to strains developed within the broader zone of deformation caused by major irregularities in the fault trace (e.g., the "Big bend" region). The Northridge earthquake was associated with movement on a blind thrust within such a zone. Another example is the strongly oblique convergent plate boundary between the Arabian and Eurasian plates where it runs through the northwestern part of the Zagros Mountains. The deformation associated with this plate boundary is partitioned into nearly pure thrust sense movements perpendicular to the boundary over a wide zone to the southwest and nearly pure strike-slip motion along the Main Recent Fault close to the actual plate boundary itself. This is demonstrated by earthquake focal mechanisms.

All tectonic plates have internal stress fields caused by their interactions with neighboring plates and sedimentary loading or unloading (e.g. deglaciation). These stresses may be sufficient to cause failure along existing fault planes, giving rise to intraplate earthquakes.

Shallow-Focus and Deep-focus Earthquakes

Collapsed Gran Hotel building in the San Salvador metropolis, after the shallow 1986 San Salvador earthquake.

The majority of tectonic earthquakes originate at the ring of fire in depths not exceeding tens of kilometers. Earthquakes occurring at a depth of less than 70 km are classified as 'shallow-focus' earthquakes, while those with a focal-depth between 70 and 300 km are commonly termed 'mid-focus' or 'intermediate-depth' earthquakes. In subduction zones, where older and colder oceanic crust descends beneath another tectonic plate, Deep-focus earthquakes may occur at much greater depths (ranging from 300 up to 700 kilometers). These seismically active areas of subduction are known as Wadati–Benioff zones. Deep-focus earthquakes occur at a depth where the subducted lithosphere should no longer be brittle, due to the high temperature and pressure. A possible mechanism for the generation of deep-focus earthquakes is faulting caused by olivine undergoing a phase transition into a spinel structure.

Earthquakes and Volcanic Activity

Earthquakes often occur in volcanic regions and are caused there, both by tectonic faults and the movement of magma in volcanoes. Such earthquakes can serve as an early warning of volcanic eruptions, as during the 1980 eruption of Mount St. Helens. Earthquake swarms can serve as markers for the location of the flowing magma throughout the volcanoes. These swarms can be recorded by seismometers and tiltmeters (a device that measures ground slope) and used as sensors to predict imminent or upcoming eruptions.

Rupture Dynamics

A tectonic earthquake begins by an initial rupture at a point on the fault surface, a process known as nucleation. The scale of the nucleation zone is uncertain, with some evidence, such as the rupture dimensions of the smallest earthquakes, suggesting that it is smaller than 100 m while other evidence, such as a slow component revealed by low-frequency spectra of some earthquakes, suggest that it is larger. The possibility that the nucleation involves some sort of preparation process is supported by the observation that about 40% of earthquakes are preceded by foreshocks. Once the rupture has initiated, it begins to propagate along the fault surface. The mechanics of this process are poorly understood, partly because it is difficult to recreate the high sliding velocities in a laboratory. Also the effects of strong ground motion make it very difficult to record information close to a nucleation zone.

Rupture propagation is generally modeled using a fracture mechanics approach, likening the rupture to a propagating mixed mode shear crack. The rupture velocity is a function of the fracture energy in the volume around the crack tip, increasing with decreasing fracture energy. The velocity of rupture propagation is orders of magnitude faster than the displacement velocity across the fault. Earthquake ruptures typically propagate at velocities that are in the range 70–90% of the S-wave velocity, and this is independent of earthquake size. A small subset of earthquake ruptures appear to have propagated at speeds greater than the S-wave velocity. These supershear earthquakes have all been observed during large strike-slip events. The unusually wide zone of coseismic damage caused by the 2001 Kunlun earthquake has been attributed to the effects of the sonic boom developed in such earthquakes. Some earthquake ruptures travel at unusually low velocities and are referred to as slow earthquakes. A particularly dangerous form of slow earthquake is the tsunami earthquake, observed where the relatively low felt intensities, caused by the slow propagation speed of some great earthquakes, fail to alert the population of the neighboring coast, as in the 1896 Sanriku earthquake.

Tidal Forces

Tides may induce some seismicity.

Earthquake Clusters

Most earthquakes form part of a sequence, related to each other in terms of location and time. Most earthquake clusters consist of small tremors that cause little to no damage, but there is a theory that earthquakes can recur in a regular pattern.

Aftershocks

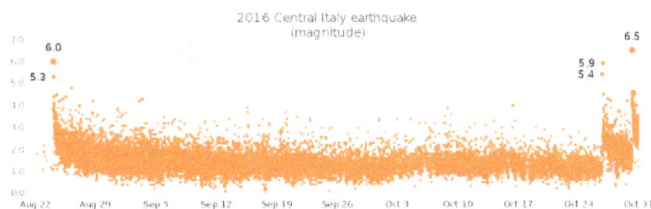

Magnitude of the Central Italy earthquakes of August and October 2016 and the aftershocks (which continued to occur after the period shown here).

An aftershock is an earthquake that occurs after a previous earthquake, the mainshock. An aftershock is in the same region of the main shock but always of a smaller magnitude. If an aftershock is larger than the main shock, the aftershock is redesignated as the main shock and the original main shock is redesignated as a foreshock. Aftershocks are formed as the crust around the displaced fault plane adjusts to the effects of the main shock.

Earthquake Swarms

Earthquake swarms are sequences of earthquakes striking in a specific area within a short period of time. They are different from earthquakes followed by a series of aftershocks by the fact that no single earthquake in the sequence is obviously the main shock, therefore none have notable higher magnitudes than the other. An example of an earthquake swarm is the 2004 activity at Yellowstone National Park. In August 2012, a swarm of earthquakes shook Southern California's Imperial Valley, showing the most recorded activity in the area since the 1970s.

Sometimes a series of earthquakes occur in what has been called an *earthquake storm*, where the earthquakes strike a fault in clusters, each triggered by the shaking or stress redistribution of the previous earthquakes. Similar to aftershocks but on adjacent segments of fault, these storms occur over the course of years, and with some of the later earthquakes as damaging as the early ones. Such a pattern was observed in the sequence of about a dozen earthquakes that struck the North Anatolian Fault in Turkey in the 20th century and has been inferred for older anomalous clusters of large earthquakes in the Middle East.

Size and Frequency of Occurrence

It is estimated that around 500,000 earthquakes occur each year, detectable with current instrumentation. About 100,000 of these can be felt. Minor earthquakes occur nearly constantly around the world in places like California and Alaska in the U.S., as well as in El Salvador, Mexico, Guatemala, Chile, Peru, Indonesia, Iran, Pakistan, the Azores in Portugal, Turkey, New Zealand, Greece, Italy, India, Nepal and Japan, but earthquakes can occur almost anywhere, including Downstate New York, England, and

Australia. Larger earthquakes occur less frequently, the relationship being exponential; for example, roughly ten times as many earthquakes larger than magnitude 4 occur in a particular time period than earthquakes larger than magnitude 5. In the (low seismicity) United Kingdom, for example, it has been calculated that the average recurrences are: an earthquake of 3.7–4.6 every year, an earthquake of 4.7–5.5 every 10 years, and an earthquake of 5.6 or larger every 100 years. This is an example of the Gutenberg–Richter law.

The Messina earthquake and tsunami took as many as 200,000 lives on December 28, 1908 in Sicily and Calabria.

The number of seismic stations has increased from about 350 in 1931 to many thousands today. As a result, many more earthquakes are reported than in the past, but this is because of the vast improvement in instrumentation, rather than an increase in the number of earthquakes. The United States Geological Survey estimates that, since 1900, there have been an average of 18 major earthquakes (magnitude 7.0–7.9) and one great earthquake (magnitude 8.0 or greater) per year, and that this average has been relatively stable. In recent years, the number of major earthquakes per year has decreased, though this is probably a statistical fluctuation rather than a systematic trend. More detailed statistics on the size and frequency of earthquakes is available from the United States Geological Survey (USGS). A recent increase in the number of major earthquakes has been noted, which could be explained by a cyclical pattern of periods of intense tectonic activity, interspersed with longer periods of low-intensity. However, accurate recordings of earthquakes only began in the early 1900s, so it is too early to categorically state that this is the case.

Most of the world's earthquakes (90%, and 81% of the largest) take place in the 40,000 km long, horseshoe-shaped zone called the circum-Pacific seismic belt, known as the Pacific Ring of Fire, which for the most part bounds the Pacific Plate. Massive earthquakes tend to occur along other plate boundaries, too, such as along the Himalayan Mountains.

With the rapid growth of mega-cities such as Mexico City, Tokyo and Tehran, in areas of high seismic risk, some seismologists are warning that a single quake may claim the lives of up to 3 million people.

Induced Seismicity

While most earthquakes are caused by movement of the Earth's tectonic plates, human activity can also produce earthquakes. Four main activities contribute to this phenomenon: storing large amounts of water behind a dam (and possibly building an extremely heavy building), drilling and injecting liquid into wells, and by coal mining and oil drilling. Perhaps the best known example is the 2008 Sichuan earthquake in China's Sichuan Province in May; this tremor resulted in 69,227 fatalities and is the 19th deadliest earthquake of all time. The Zipingpu Dam is believed to have fluctuated the pressure of the fault 1,650 feet (503 m) away; this pressure probably increased the power of the earthquake and accelerated the rate of movement for the fault. The greatest earthquake in Australia's history is also claimed to be induced by humanity, through coal mining. The city of Newcastle was built over a large sector of coal mining areas. The earthquake has been reported to be spawned from a fault that reactivated due to the millions of tonnes of rock removed in the mining process.

Measuring and Locating Earthquakes

Earthquakes can be recorded by seismometers up to great distances, because seismic waves travel through the whole Earth's interior. The absolute magnitude of a quake is conventionally reported by numbers on the moment magnitude scale (formerly Richter scale, magnitude 7 causing serious damage over large areas), whereas the felt magnitude is reported using the modified Mercalli intensity scale (intensity II–XII).

Every tremor produces different types of seismic waves, which travel through rock with different velocities:

- Longitudinal P-waves (shock- or pressure waves)

- Transverse S-waves (both body waves)

- Surface waves — (Rayleigh and Love waves)

Propagation velocity of the seismic waves ranges from approx. 3 km/s up to 13 km/s, depending on the density and elasticity of the medium. In the Earth's interior the shock- or P waves travel much faster than the S waves (approx. relation 1.7 : 1). The differences in travel time from the epicenter to the observatory are a measure of the distance and can be used to image both sources of quakes and structures within the Earth. Also the depth of the hypocenter can be computed roughly.

In solid rock P-waves travel at about 6 to 7 km per second; the velocity increases within the deep mantle to ~13 km/s. The velocity of S-waves ranges from 2–3 km/s in light sediments and 4–5 km/s in the Earth's crust up to 7 km/s in the deep mantle. As a consequence, the first waves of a distant earthquake arrive at an observatory via the Earth's mantle.

On average, the kilometer distance to the earthquake is the number of seconds between the P and S wave times 8. Slight deviations are caused by inhomogeneities of subsurface structure. By such analyses of seismograms the Earth's core was located in 1913 by Beno Gutenberg.

Earthquakes are not only categorized by their magnitude but also by the place where they occur. The world is divided into 754 Flinn–Engdahl regions (F-E regions), which are based on political and geographical boundaries as well as seismic activity. More active zones are divided into smaller F-E regions whereas less active zones belong to larger F-E regions.

Standard reporting of earthquakes includes its magnitude, date and time of occurrence, geographic coordinates of its epicenter, depth of the epicenter, geographical region, distances to population centers, location uncertainty, a number of parameters that are included in USGS earthquake reports (number of stations reporting, number of observations, etc.), and a unique event ID.

Effects of Earthquakes

1755 copper engraving depicting Lisbon in ruins and in flames after the 1755 Lisbon earthquake, which killed an estimated 60,000 people. A tsunami overwhelms the ships in the harbor.

The effects of earthquakes include, but are not limited to, the following:

Shaking and Ground Rupture

Shaking and ground rupture are the main effects created by earthquakes, principally resulting in more or less severe damage to buildings and other rigid structures. The severity of the local effects depends on the complex combination of the earthquake magnitude, the distance from the epicenter, and the local geological and geomorphological conditions, which may amplify or reduce wave propagation. The ground-shaking is measured by ground acceleration.

Damaged buildings in Port-au-Prince, Haiti, January 2010.

Specific local geological, geomorphological, and geostructural features can induce high levels of shaking on the ground surface even from low-intensity earthquakes. This effect is called site or local amplification. It is principally due to the transfer of the seismic motion from hard deep soils to soft superficial soils and to effects of seismic energy focalization owing to typical geometrical setting of the deposits.

Ground rupture is a visible breaking and displacement of the Earth's surface along the trace of the fault, which may be of the order of several meters in the case of major earthquakes. Ground rupture is a major risk for large engineering structures such as dams, bridges and nuclear power stations and requires careful mapping of existing faults to identify any which are likely to break the ground surface within the life of the structure.

Landslides and Avalanches

Earthquakes, along with severe storms, volcanic activity, coastal wave attack, and wildfires, can produce slope instability leading to landslides, a major geological hazard. Landslide danger may persist while emergency personnel are attempting rescue.

Fires

Fires of the 1906 San Francisco earthquake

Earthquakes can cause fires by damaging electrical power or gas lines. In the event of water mains rupturing and a loss of pressure, it may also become difficult to stop the spread of a fire once it has started. For example, more deaths in the 1906 San Francisco earthquake were caused by fire than by the earthquake itself.

Soil Liquefaction

Soil liquefaction occurs when, because of the shaking, water-saturated granular material (such as sand) temporarily loses its strength and transforms from a solid to a liquid. Soil liquefaction may cause rigid structures, like buildings and bridges, to tilt or sink into the liquefied deposits. For example, in the 1964 Alaska earthquake, soil liquefaction caused many buildings to sink into the ground, eventually collapsing upon themselves.

Tsunami

The tsunami of the 2004 Indian Ocean earthquake

Tsunamis are long-wavelength, long-period sea waves produced by the sudden or abrupt movement of large volumes of water. In the open ocean the distance between wave crests can surpass 100 kilometers (62 mi), and the wave periods can vary from five minutes to one hour. Such tsunamis travel 600-800 kilometers per hour (373–497 miles per hour), depending on water depth. Large waves produced by an earthquake or a submarine landslide can overrun nearby coastal areas in a matter of minutes. Tsunamis can also travel thousands of kilometers across open ocean and wreak destruction on far shores hours after the earthquake that generated them.

Ordinarily, subduction earthquakes under magnitude 7.5 on the Richter scale do not cause tsunamis, although some instances of this have been recorded. Most destructive tsunamis are caused by earthquakes of magnitude 7.5 or more.

Floods

A flood is an overflow of any amount of water that reaches land. Floods occur usually when the volume of water within a body of water, such as a river or lake, exceeds the total capacity of the formation, and as a result some of the water flows or sits outside of the normal perimeter of the body. However, floods may be secondary effects of earthquakes, if dams are damaged. Earthquakes may cause landslips to dam rivers, which collapse and cause floods.

The terrain below the Sarez Lake in Tajikistan is in danger of catastrophic flood if the landslide dam formed by the earthquake, known as the Usoi Dam, were to fail during a future earthquake. Impact projections suggest the flood could affect roughly 5 million people.

Human Impacts

Ruins of the Għajn Ħadid Tower, which collapsed in an earthquake in 1856

An earthquake may cause injury and loss of life, road and bridge damage, general property damage, and collapse or destabilization (potentially leading to future collapse) of buildings. The aftermath may bring disease, lack of basic necessities, mental consequences such as panic attacks, depression to survivors, and higher insurance premiums.

Major Earthquakes

One of the most devastating earthquakes in recorded history was the 1556 Shaanxi earthquake, which occurred on 23 January 1556 in Shaanxi province, China. More than 830,000 people died. Most houses in the area were yaodongs—dwellings carved out of loess hillsides—and many victims were killed when these structures collapsed. The 1976 Tangshan earthquake, which killed between 240,000 and 655,000 people, was the deadliest of the 20th century.

The 1960 Chilean earthquake is the largest earthquake that has been measured on a seismograph, reaching 9.5 magnitude on 22 May 1960. Its epicenter was near Cañete, Chile. The energy released was approximately twice that of the next most powerful earthquake, the Good Friday earthquake (March 27, 1964) which was centered in Prince William Sound, Alaska. The ten largest recorded earthquakes have all been megathrust earthquakes; however, of these ten, only the 2004 Indian Ocean earthquake is simultaneously one of the deadliest earthquakes in history.

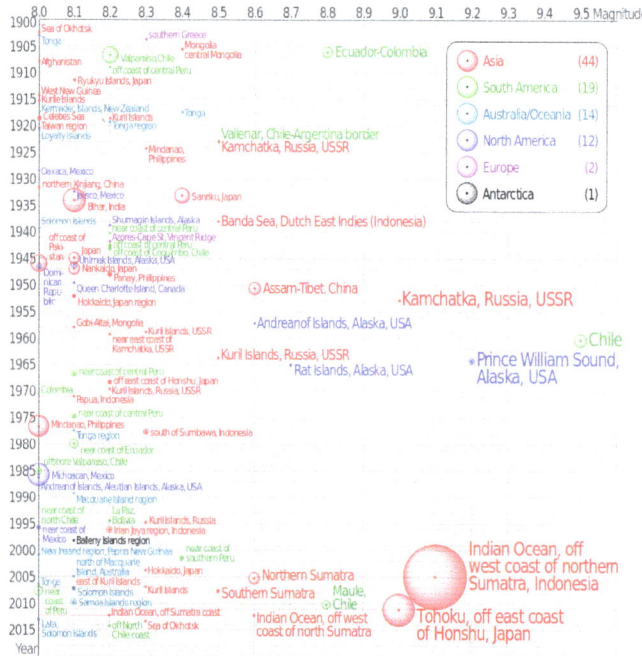

Earthquakes of magnitude 8.0 and greater since 1900. The apparent 3D volumes of the bubbles are linearly proportional to their respective fatalities.

Earthquakes that caused the greatest loss of life, while powerful, were deadly because of their proximity to either heavily populated areas or the ocean, where earthquakes often create tsunamis that can devastate communities thousands of kilometers away. Regions most at risk for great loss of life include those where earthquakes are relatively rare but powerful, and poor regions with lax, unenforced, or nonexistent seismic building codes.

Prediction

Many methods have been developed for predicting the time and place in which earthquakes will occur. Despite considerable research efforts by seismologists, scientifically reproducible predictions cannot yet be made to a specific day or month. However, for well-understood faults the probability that a segment may rupture during the next few decades can be estimated.

Earthquake warning systems have been developed that can provide regional notification of an earthquake in progress, but before the ground surface has begun to move, potentially allowing people within the system's range to seek shelter before the earthquake's impact is felt.

Preparedness

The objective of earthquake engineering is to foresee the impact of earthquakes on buildings and other structures and to design such structures to minimize the risk of damage. Existing structures can be modified by seismic retrofitting to improve their resistance to earthquakes. Earthquake insurance can provide building owners with financial protection against losses resulting from earthquakes.

Emergency management strategies can be employed by a government or organization to mitigate risks and prepare for consequences.

Historical Views

Tremblement de terre en Italie. 340 ans avant J.-C. — L. Papirius Cursor consul (d'après Lycosthène).

An image from a 1557 book

From the lifetime of the Greek philosopher Anaxagoras in the 5th century BCE to the 14th century CE, earthquakes were usually attributed to "air (vapors) in the cavities of the Earth." Thales of Miletus, who lived from 625–547 (BCE) was the only documented person who believed that earthquakes were caused by tension between the earth and water. Other theories existed, including the Greek philosopher Anaxamines' (585–526 BCE) beliefs that short incline episodes of dryness and wetness caused seismic activity. The Greek philosopher Democritus (460–371 BCE) blamed water in general for earthquakes. Pliny the Elder called earthquakes "underground thunderstorms."

Recent Studies

In recent studies, geologists claim that global warming is one of the reasons for increased seismic activity. According to these studies melting glaciers and rising sea levels disturb the balance of pressure on Earth's tectonic plates thus causing increase in the frequency and intensity of earthquakes.

Earthquakes in Culture

Mythology and Religion

In Norse mythology, earthquakes were explained as the violent struggling of the god Loki. When Loki, god of mischief and strife, murdered Baldr, god of beauty and light, he was punished by being bound in a cave with a poisonous serpent placed above his head dripping venom. Loki's wife Sigyn stood by him with a bowl to catch the poison, but whenever she had to empty the bowl the poison dripped on Loki's face, forcing him to jerk his head away and thrash against his bonds, which caused the earth to tremble.

In Greek mythology, Poseidon was the cause and god of earthquakes. When he was in a bad mood, he struck the ground with a trident, causing earthquakes and other calamities. He also used earthquakes to punish and inflict fear upon people as revenge.

In Japanese mythology, Namazu (鯰) is a giant catfish who causes earthquakes. Namazu lives in the mud beneath the earth, and is guarded by the god Kashima who restrains the fish with a stone. When Kashima lets his guard fall, Namazu thrashes about, causing violent earthquakes.

In Popular Culture

In modern popular culture, the portrayal of earthquakes is shaped by the memory of great cities laid waste, such as Kobe in 1995 or San Francisco in 1906. Fictional earthquakes tend to strike suddenly and without warning. For this reason, stories about earthquakes generally begin with the disaster and focus on its immediate aftermath, as in *Short Walk to Daylight* (1972), *The Ragged Edge* (1968) or *Aftershock: Earthquake in New York* (1999). A notable example is Heinrich von Kleist's classic novella, *The Earthquake in Chile*, which describes the destruction of Santiago in 1647. Haruki Murakami's short fiction collection After the Quake depicts the consequences of the Kobe earthquake of 1995.

The most popular single earthquake in fiction is the hypothetical "Big One" expected of California's San Andreas Fault someday, as depicted in the novels *Richter 10* (1996), *Goodbye California* (1977), *2012* (2009) and *San Andreas* (2015) among other works. Jacob M. Appel's widely anthologized short story, *A Comparative Seismology*, features a con artist who convinces an elderly woman that an apocalyptic earthquake is imminent.

Contemporary depictions of earthquakes in film are variable in the manner in which they reflect human psychological reactions to the actual trauma that can be caused to directly afflicted families and their loved ones. Disaster mental health response research emphasizes the need to be aware of the different roles of loss of family and key community members, loss of home and familiar surroundings, loss of essential supplies and services to maintain survival. Particularly for children, the clear availability of caregiving adults who are able to protect, nourish, and clothe them in the aftermath of the earthquake, and to help them make sense of what has befallen them has been shown even more important to their emotional and physical health than the simple giving of provisions. As was observed after other disasters involving destruction and loss of life and their media depictions, recently observed in the 2010 Haiti earthquake, it is also important not to pathologize the reactions to loss and displacement or disruption of governmental administration and services, but rather to validate these reactions, to support constructive problem-solving and reflection as to how one might improve the conditions of those affected.

Seismotectonics

Seismotectonics is the study of the relationship between the earthquakes, active tectonics and individual faults of a region. It seeks to understand which faults are responsible for seismic activity in an area by analysing a combination of regional tectonics, recent instrumentally recorded events, accounts of historical earthquakes and geomorphological evidence. This information can then be used to quantify the seismic hazard of an area.

Methodology

A seismotectonic analysis of an area often involves the integration of disparate datasets.

Regional Tectonics

An understanding of the regional tectonics of an area is likely to be derived from published geological maps, research publications on the geological structure and seismic reflection profiles, where available, augmented by other geophysical data.

In order to understand the seismic hazard of an area it is necessary not only to know where potentially active faults are, but also the orientation of the stress field. This is normally derived from a combination of earthquake data, borehole breakout analysis, direct stress measurement and the analysis of geologically young fault networks. The World Stress Map Project provides a useful online compilation of such data.

Earthquakes

Instrumentally Recorded Events

Since the early 20th century, sufficient information has been available from seismometers to allow the location, depth and magnitude of earthquakes to be calculated. In terms of identifying the fault responsible for an earthquake where there is no clear surface trace, recording the locations of aftershocks generally gives a strong indication of the strike of the fault.

In the last 30 years, it has been possible to routinely calculate focal mechanisms from teleseismic data. Catalogues of events with calculated focal mechanisms are now available online, such as the searchable catalogue from the NEIC. As focal mechanisms give two potential active fault plane orientations, other evidence is required to interpret the origin of an individual event. Although only available for a restricted time period, in areas of moderate to intense seismicity there is probably sufficient data to characterise the type of seismicity in an area, if not all the active structures.

Historical Records

Attempts to understand the seismicity of an area require information from earthquakes before the era of instrumental recording.[viii] This requires a careful assessment of historical data in terms of their reliability. In most cases, all that can be derived is an estimate of the location and magnitude of the event. However, such data is needed to fill the gaps in the instrumental record, particularly in areas with either relatively low seismicity or where the repeat periods for major earthquakes is more than a hundred years.

Field Investigations

Information on the timing and magnitude of seismic events that occurred before instrumental recording can be obtained from excavations across faults that are thought to be seismically active and by studying recent sedimentary sequences for evidence of seismic activity such as seismites or tsunami deposits.

Geomorphology

Seismically active faults and related fault generated folds have a direct effect on the geomorphology of a region. This may allow the direct identification of active structures not previously known. In some cases such observations can be used quantitatively to constrain the repeat period of major earthquakes, such as the raised beaches of Turakirae Head recording the history of coseismic uplift of the Rimutaka Range due to displacement on the Wairarapa Fault in North Island, New Zealand.

Intraplate Earthquake

An intraplate earthquake occurs in the interior of a tectonic plate, whereas an interplate earthquake is one that occurs at a plate boundary.

Intraplate earthquakes are relatively rare. Interplate earthquakes, which occur at plate boundaries, are more common. Nonetheless, very large intraplate earthquakes can inflict heavy damage, particularly because such areas are not accustomed to earthquakes and buildings are usually not seismically retrofitted. Examples of damaging intraplate earthquakes are the devastating Gujarat earthquake in 2001, the 2012 Indian Ocean earthquakes, the 1811-1812 earthquakes in New Madrid, Missouri, and the 1886 earthquake in Charleston, South Carolina.

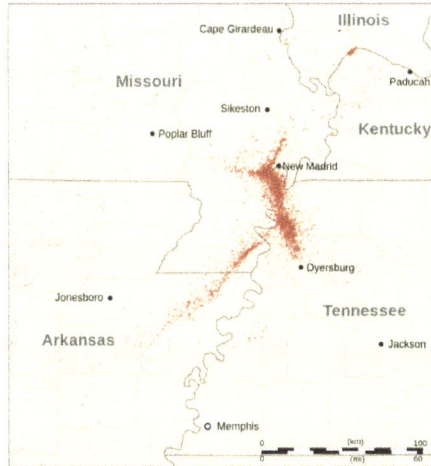

Distribution of seismicity associated with the New Madrid Seismic Zone (since 1974). This zone of intense earthquake activity is located deep in the interior of the North American plate.

Fault Zones within Tectonic Plates

The surface of the Earth is made up of seven primary and eight secondary tectonic plates, plus dozens of tertiary microplates. The large plates move very slowly, owing to convection currents within the mantle below the crust. Because they do not all move in the same direction, plates often directly collide or move laterally along each other, a tectonic environment that makes earthquakes frequent. Relatively few earthquakes occur in intraplate environments; most occur on faults near plate margins. By definition, intraplate earthquakes do not occur near plate boundaries, but along faults in the normally stable interior of plates. These earthquakes often occur at the location of ancient failed rifts, because such old structures may present a weakness in the crust where it can easily slip to accommodate regional tectonic strain.

Compared to earthquakes near plate boundaries, intraplate earthquakes are not well understood, and the hazards associated with them may be difficult to quantify.

Historic Examples

Historic examples of intraplate earthquakes include those in Mineral, Virginia in 2011 (estimated magnitude 5.8), New Madrid in 1811 and 1812 (estimated magnitude as high as 8.1), the Boston (Cape Ann) earthquake of 1755 (estimated magnitude 6.0 to 6.3), earthquakes felt in New York City in 1737 and 1884 (both quakes estimated at about

5.5 magnitude), and the Charleston earthquake in South Carolina in 1886 (estimated magnitude 6.5 to 7.3). The Charleston quake was particularly surprising because, unlike Boston and New York, the area had almost no history of even minor earthquakes.

In 2001, a large intraplate earthquake devastated the region of Gujarat, India. The earthquake occurred far from any plate boundaries, which meant the region above the epicenter was unprepared for earthquakes. In particular, the Kutch district suffered tremendous damage, where the death toll was over 12,000 and the total death toll was higher than 20,000.

Causes

Many cities live with the seismic risk of a rare, large intraplate earthquake. The cause of these earthquakes is often uncertain. In many cases, the causative fault is deeply buried, and sometimes cannot even be found. Under these circumstances it is difficult to calculate the exact seismic hazard for a given city, especially if there was only one earthquake in historical times. Some progress is being made in understanding the fault mechanics driving these earthquakes.

Prediction

Scientists continue to search for the causes of these earthquakes, and especially for some indication of how often they recur. The best success has come with detailed micro-seismic monitoring, involving dense arrays of seismometers. In this manner, very small earthquakes associated with a causative fault can be located with great accuracy, and in most cases these line up in patterns consistent with faulting. Cryoseisms can sometimes be mistaken for intraplate earthquakes.

Interplate Earthquake

An interplate earthquake is an earthquake that occurs at the boundary between two tectonic plates. Earthquakes of this type account for more than 90 percent of the total seismic energy released around the world. If one plate is trying to move past the other, they will be locked until sufficient stress builds up to cause the plates to slip relative to each other. The slipping process creates an earthquake with land deformations and resulting seismic waves which travel through the Earth and along the Earth's surface. Relative plate motion can be lateral as along a transform fault boundary or vertical if along a convergent subduction boundary or a rift at a divergent boundary. At a subduction boundary the motion is due to one plate slipping beneath the other plate resulting in an interplate thrust or megathrust earthquake, which are the most powerful earthquakes.

Some areas of the world that are particularly prone to such events include the west coast of North America (especially California and Alaska), the northeastern Mediterranean region (Greece, Italy, and Turkey in particular), Iran, New Zealand, Indonesia, India, Japan, and parts of China.

Interplate earthquakes differ from intraplate earthquake in the intensity of stress drop which occurs after the quake. Intraplate earthquake have, on average, more stress drop than that of the interplate earthquake. Interplate earthquakes also differ fundamentally from intraplate earthquakes in the way stress is released and recovered. An interplate earthquake results in an immediate stress drop along the fault. Following this is a period of postseismic stress restoration. This restoration occurs quickly within the first few decades following the rupture and is due to tectonic loading and viscous relaxation in the lower crust. This results in a transfer of stress to the upper crust. Later on, a period of steady stress increase occurs due to tectonic loading.

References

- Ohnaka, M. (2013). The Physics of Rock Failure and Earthquakes. Cambridge University Press. p. 148. ISBN 9781107355330.

- George E. Dimock (1990). The Unity of the Odyssey. Univ of Massachusetts Press. pp. 179–. ISBN 0-87023-721-7.

- Van Riper, A. Bowdoin (2002). Science in popular culture: a reference guide. Westport: Greenwood Press. p. 60. ISBN 0-313-31822-0.

- Ambraseys, Nicolas; Melville, C.P. (1982). A History of Persian Earthquakes (PDF). Cambridge University Press. ISBN 9780521021876.

- Bolt, Bruce (August 2005), Earthquakes: 2006 Centennial Update – The 1906 Big One (Fifth ed.), W. H. Freeman and Company, p. 150, ISBN 978-0716775485

- "Fire and Ice: Melting Glaciers Trigger Earthquakes, Tsunamis and Volcanos". about News. Retrieved October 27, 2015.

- Geographic.org. "Magnitude 8.0 - SANTA CRUZ ISLANDS Earthquake Details". Gobal Earthquake Epicenters with Maps. Retrieved 2013-03-13.

- "On Shaky Ground, Association of Bay Area Governments, San Francisco, reports 1995,1998 (updated 2003)". Abag.ca.gov. Retrieved 2010-08-23.

- Kanamori Hiroo. "The Energy Release in Great Earthquakes" (PDF). Journal of Geophysical Research. Retrieved 2010-10-10.

- Pressler, Margaret Webb (14 April 2010). "More earthquakes than usual? Not really.". KidsPost. Washington Post: Washington Post. pp. C10.

- Hjaltadóttir S., 2010, "Use of relatively located microearthquakes to map fault patterns and estimate the thickness of the brittle crust in Southwest Iceland"

Permissions

All chapters in this book are published with permission under the Creative Commons Attribution Share Alike License or equivalent. Every chapter published in this book has been scrutinized by our experts. Their significance has been extensively debated. The topics covered herein carry significant information for a comprehensive understanding. They may even be implemented as practical applications or may be referred to as a beginning point for further studies.

We would like to thank the editorial team for lending their expertise to make the book truly unique. They have played a crucial role in the development of this book. Without their invaluable contributions this book wouldn't have been possible. They have made vital efforts to compile up to date information on the varied aspects of this subject to make this book a valuable addition to the collection of many professionals and students.

This book was conceptualized with the vision of imparting up-to-date and integrated information in this field. To ensure the same, a matchless editorial board was set up. Every individual on the board went through rigorous rounds of assessment to prove their worth. After which they invested a large part of their time researching and compiling the most relevant data for our readers.

The editorial board has been involved in producing this book since its inception. They have spent rigorous hours researching and exploring the diverse topics which have resulted in the successful publishing of this book. They have passed on their knowledge of decades through this book. To expedite this challenging task, the publisher supported the team at every step. A small team of assistant editors was also appointed to further simplify the editing procedure and attain best results for the readers.

Apart from the editorial board, the designing team has also invested a significant amount of their time in understanding the subject and creating the most relevant covers. They scrutinized every image to scout for the most suitable representation of the subject and create an appropriate cover for the book.

The publishing team has been an ardent support to the editorial, designing and production team. Their endless efforts to recruit the best for this project, has resulted in the accomplishment of this book. They are a veteran in the field of academics and their pool of knowledge is as vast as their experience in printing. Their expertise and guidance has proved useful at every step. Their uncompromising quality standards have made this book an exceptional effort. Their encouragement from time to time has been an inspiration for everyone.

The publisher and the editorial board hope that this book will prove to be a valuable piece of knowledge for students, practitioners and scholars across the globe.

Index